MICHEL LEBOEUF

THE LOST SONGS OF NATURE

Nature's Symphony in the Age of Noise Pollution

translated by Neil Macmillan

Other books by Michel Leboeuf

Éditions MultiMondes
Le dernier caribou, 2020 (shortlisted, Prix Hubert-Reeves, 2021)
Paroles d'un bouleau jaune,
2018 *Arbres en lumière*, 2016

Éditions Michel Quintin
Papillons de nuit et chenilles du Québec et des Maritimes, 2018
Papillons et chenilles du Québec et des Maritimes, 2012
Famille nature : Jouer dehors au Québec, 2008
Arbres et plantes forestières du Québec et des Maritimes, 2007
Découvrir les oiseaux du Québec et des Maritimes, 2005

Éditions Orinha Media
Le Québec en miettes, 2012 (Prix Hubert-Reeves, 2013)
Nous n'irons plus au bois, 2010 (Prix Hubert-Reeves, 2011)

MICHEL LEBOEUF

LOST SONGS OF NATURE

NATURE'S SYMPHONY IN THE AGE OF NOISE POLLUTION

GREAT PLAINS PRESS

Great Plains Publications
320 Rosedale Avenue
Winnipeg, MB R3G 1H1
www.greatplainspress.ca

Great Plains Publications gratefully acknowledges the financial support provided for its publishing program by the Government of Canada through the Canada Book Fund; the Canada Council for the Arts; the Province of Manitoba through the Book Publishing Tax Credit and the Book Publisher Marketing Assistance Program; and the Manitoba Arts Council.

The translation of this work was made possible thanks to the financial support of the Société de développement des entreprises culturelles du Québec (SODEC).

Translation by Neil Macmillan
Design & Typography by Electric Monk Media Ltd.
Printed in Canada by Friesens

Original title in French: ***Les chants perdus de la nature : Bienvenue en Anthropophonie***
Written by Michel Leboeuf

Library and Archives Canada Cataloguing in Publication

Title: Lost songs of nature : nature's symphony in the age of noise pollution / Michel Leboeuf ; translated by Neil Macmillan.
Other titles: Chants perdus de la nature. English
Names: Lebœuf, Michel, 1962- author
Description: Translation of: Les chants perdus de la nature : bienvenue en anthropophonie. | Includes bibliographical references and index.
Identifiers: Canadiana (print) 20250178257 | Canadiana (ebook) 20250178281 | ISBN 9781773371351 (softcover) | ISBN 9781773371368 (EPUB)
Subjects: LCSH: Nature sounds—Environmental aspects. | LCSH: Bioacoustics—Environmental aspects.
Classification: LCC QH510.5 .L4313 2025 | DDC 591.59/4—dc23

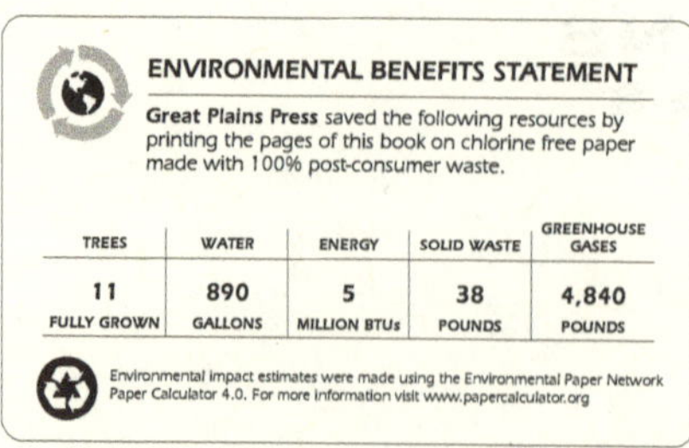

ENVIRONMENTAL BENEFITS STATEMENT

Great Plains Press saved the following resources by printing the pages of this book on chlorine free paper made with 100% post-consumer waste.

TREES	WATER	ENERGY	SOLID WASTE	GREENHOUSE GASES
11	890	5	38	4,840
FULLY GROWN	GALLONS	MILLION BTUs	POUNDS	POUNDS

Environmental impact estimates were made using the Environmental Paper Network Paper Calculator 4.0. For more information visit www.papercalculator.org

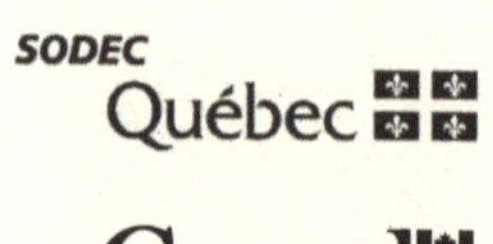

Canada

TABLE OF CONTENTS

FOREWORD

The Lost Songs of Nature is a paean to ecology and the beauty of our planet. It's also a call to open our eyes, our ears, and our minds. As author and ornithologist Michel Leboeuf says: "We are rapidly destroying the Earth's biodiversity that has taken almost 4 billion years to form."

The human species—or, as it likes to call itself, *Homo sapiens*, "the wise person"—is a highly visual organism. Our visual acuity is among the most developed in the animal world, behind only a few birds of prey. However, thanks to our ability to communicate through sound and speech, we have been able to transmit and store our culture and knowledge down through the ages. This human heritage has been passed on orally from generation to generation. What's more, modern humans have become decidedly vocal beings.

In this book, Leboeuf presents a fresh and original conception of our world in five symphonic movements stretching from the last ice age to modern times. Imagine an organism that has lived for the last 10,000 years without eyes to see or a nose to smell, but with prodigious hearing. What would it say about this dynamic and vibrant natural world we have here on Earth? What would it think of us humans? That we're resilient, creative, and capable of change, but also boorish, excessively consumptive, and extremely noisy!

In this book, we can sense Leboeuf's passion for Quebec's flora and fauna, and how this natural life interacts with its environment. We can easily imagine his ornithologist's attentive ear and biologist's humility in the face of nature's magnificent living orchestra. The academic in me would like to highlight the in-depth documentary research underpinning Leboeuf's text, which is based on the latest scientific knowledge. In an impressive demonstration of independent learning about biodiversity, the book includes more than 170 references to animal and plant species.

Some of my own research in conservation biology deals with soundscapes. In that regard, I particularly appreciate how Leboeuf invites his readers to discover the diversity of modes of acoustic communication that exists throughout the Tree of Life. This diversity includes the *quaa* alarm call of eastern grey squirrels, the *coo-woo* serenade of coquettish grouse, and the *claps!* of the tails of American beavers (as they're officially called) on the surface of water. The chapter on plant soundscapes is particularly fascinating, and admittedly made me rethink my previous conceptions. Indeed, I was seized with an irresistible urge to assess my own plants' learning capacities, and of course, listen to the songs of the trees.

Soundscape ecology is a fast-growing field. We continue to be amazed by the striking variety of sounds produced by fish underwater, or the acoustic ambience of the ground beneath our feet. Every day, we discover a little more about the impact of all the sounds generated by our extremely noisy species.

In a single paragraph, Leboeuf sums up how our homogenization of nature is a symptom of our insatiable appetite for converting nature into cash. When we plant corn and soybean monocultures on 80% of our arable land, that's a problem. When the only sound entering our homes is that of vehicle traffic outside, that's a problem. When a teenager's acoustic environment is dominated by the sound of videogame gunfire, that's a problem. When biodiversity has all but disappeared, that's a problem. Instead, why not make all these annoying environments a little richer and more interesting? That would be a step in the right direction.

This book is full of examples that make it easy for neophytes to understand the relationship between biodiversity and acoustic ecology. Even if the underlying message is occasionally pessimistic, the author emphasizes solutions. We know what they are and can implement them together. It is now more necessary than ever to remain optimistic about conserving nature. All too often we've tried to frighten people into saving the planet. Of course, fear is a powerful motivator, providing that it really scares people, otherwise results will be slow in coming. The human species must find and accept its role and position on the stage of life. We need to stop being rebellious teenagers and become more mature. Let's be clear-sighted and optimistic at the same time!

Raphaël Proulx
Director, Centre de recherche sur les interactions bassins versants - écosystèmes aquatiques
Professor of Conservation Biology
Université du Québec à Trois-Rivières

PRELUDE

All too often we only becomes interested in something when we're about to lose it.

I find that my hearing is steadily becoming less sharp, accurate, and reliable as I get older. I find that I can hear fewer and fewer high-frequency sounds, such as the sharp-pitched songs of some of the forest's most colourful passerines like warblers or kinglets.

I've often spent many hours listening to birds, not only for pleasure, but also as part of my work inventorying birds in the spring, when species are counted—mainly by ear— in specific areas and over specific time periods. When I started to lose my ability to catch frequencies approaching the upper limits of human hearing, i.e., 20,000 hertz (Hz), I had to resort to technology—a high-quality sound recorder—to continue doing this work. Once back home, I could then turn my headphones up to maximum volume and compare my handwritten notes with the audio files. I invariably noticed that several forest species with the highest-pitched songs, such as the black-and-white warbler, the golden-crowned kinglet, and the brown creeper, were there in the field during the counting period but I hadn't realized it.

It was only with my headphones on that I discovered (or rather, rediscovered) all the rich sounds of those spring days in the deep forest.

I found myself once again marvelling at nature's symphony, just as I did when I was eight or nine years old. On the other hand, I also became concerned about how the natural soundscapes in my part of the world were being obliterated. And rightly so, because on almost every track I recorded, it was impossible to have ten minutes of natural ambient sound without hearing, at regular intervals, man-made sounds like a jumbo jet in the sky, a motorbike on a faraway road, or a chainsaw in a nearby forest.

✦

Living organisms use different senses to obtain information about their environment. Vertebrates—fish, reptiles, birds, mammals and so on—possess a complex central nervous system that continually analyzes an enormous quantity of visual, olfactory, tactile, and other phenomena.

Other animals use a variety of organs or structures, such as the feathery antennae of moths or the cilia of paramecia (tiny single-celled freshwater organisms), to obtain information to enable them to move about and feed.

Even plants have developed mechanisms to gather information about their environment and react as needed. Two well-known examples of this phenomenon are phototrophism (plants' turning towards the light) and root tactility (the ability of plants' adventitious roots to continue growing by detecting and bypassing underground obstacles).

In humanity's distant past, one of our senses—sight—ended up dominating the other four, especially from the time when we started to become a more sedentary species. With the advent of agriculture some 10,000 years ago, we also had to invent a way of recording information on how to use agrarian technologies, how to construct buildings, how to organize civil society, and even how to wage war with neighbouring cities or states. As the saying goes, "necessity is the mother of invention," and so we began to write!

Since then, sight has reigned supreme over our sensory environment, and this domination only accentuated during the Age of Enlightenment, the historical turning point in the seventeenth and eighteenth centuries when science and technology basically took over society. In the past, writing made it possible to transmit knowledge in all fields, such as biology and philosophy, from one generation to the next. However, whereas knowledge used to be transmitted via paper, it is now transmitted digitally with only traditional societies continuing to use oral transmission as their principal means of cultural dissemination.

The supremacy of sight over the other senses, which is a marked feature of contemporary life, is evident in such everyday expressions as "I see what you mean" and "You have to see it to believe it." In fact, when we say, "it's blindingly obvious," we actually mean you don't have to see it to believe it!

On the other hand, the natural world needs to be not only observed but also heard.

That is why we must also become interested in the rich and diverse soundscapes of the world around us.

✦

Canadian composer, theorist, and ecologist R. Murray Schafer (1933–2021) is credited with formulating the concept of *soundscape*, the neologism he proposed in 1977 by analogy with the term *landscape*, which refers to the visual environment.

Instead of elements that can be observed with the eyes, soundscapes can be listened to. Soundscapes, consisting of sounds and noises, make up immersive acoustic environments for those who pay attention to them, and can be natural (for example, the sound of waves on the shore, or the songs of birds) or generated by human activity (the roar of traffic or the ringtones of a phone).

Like visual landscapes, soundscapes consist of disparate sets of elements in the broadest panoramic perception possible—on all sides and at all distances—in relation to the human perceiver.

To study the soundscapes in which humans and all the other living organisms with which we share the planet are immersed, a new discipline has been invented: *acoustic ecology* (or sound studies).

Acoustic ecology is thus nothing more or less than the study of the *sonosphere*[1]—the 'music of the world.'

For Schafer, the universe is one gigantic musical composition, with no beginning and no end. That is why he opens his seminal work, *The Tuning of the World*, by quoting a definition of music by experimental contemporary music composer John Cage: "Music is sounds, sounds around us, whether we're in or out of concert halls."

In this respect, one of John Cage's most singular works is undoubtedly *4′33″*, a piece in which an instrumentalist or entire orchestra "interprets" a stretch of silence for precisely four minutes and thirty-three seconds. The aim of the piece is to capture the ambient noises that inevitably occur during a concert. By the same token, Cage highlighted the ideas of poet and naturalist Henry David Thoreau (1817-62), who observed in his book *Walden* that it was much more interesting to listen to the sounds of Mother Nature—the sounds generated by animals or natural phenomena, such as the wind in the leaves—than to listen to the premeditated, formatted forms of music created by composers for human audiences.

✦

The Lost Songs of Nature is an invitation to listen, discover, and rediscover the music of ecosystems—the music of forests, marshes, swamps, bogs and seashores—and thus, as discussed in this book's final chapters, take an interest in all the conservation measures that can be taken to keep these ecosystems as rich and admirable as they inherently are.

The Lost Songs of Nature is divided into five sections or *movements*,

[1] A neologism analogous to biosphere, suggested by ecoacoustics researcher Jérôme Sueur

as in an orchestral score.

The first movement, *And Sound Came to Upper America*, is impressionistic: it invites the reader to step back in time to the last ice age and imagine the successive soundscapes that followed each other in northeastern Upper America when we humans were not yet around.

The second movement, *The Nature of Sound*, is somewhat more technical. It covers the main classes of sound and explains the physical factors that make up sound.

The third movement, *Animal Soundscapes*, deals with how animals hear and how they employ the calls, songs, and other noises they produce during the various stages of their lives.

The fourth movement, *Plant Soundscapes*, addresses the sensory perception of plants, and, yes, their ability to hear sounds. This section describes the advances made in a flourishing and totally new field of scientific investigation: plant bioacoustics.

The fifth and final movement, *The Songs of the Trees*, shows that humans, as members of the planet's living orchestra, have become musicians who play too loud and no longer follow the score. In conclusion, I make a few suggestions on how we humans can scale back our stranglehold over everything.

✦

The common loon on the lake falls silent when the seaplane lands; the torrent's murmur is drowned out by the roar of vehicle traffic; and the wolf up there, on the edge of the boreal forest, stops its howling when the long lumber truck convoys head south to the sawmills.

Every day, the symphony of natural life gets quieter, losing texture and richness in proportion to each square metre of the planet's surface encroached upon by humans.

That's why it's so important to learn about these natural soundscapes before they're lost forever!

First Movement

And Sound Came to Upper America

And the land under my feet imbibes the silence.

Natasha Kanapé Fontaine
Manifeste Assi

1

A Brief History of Quebec Soundscapes

Approximately 12,500 years ago, somewhere on the north shore of the Champlain Sea, a vast expanse of water is slowly becoming less salty. It's a body of water that the Innu and Abenaki people in turn would later call *Wepistukujaw Sipo* and *Moliantegok*, and which colonists from Europe would subsequently name the St. Lawrence River.

Everything back then was covered by an enormous ice cap several hundred metres high: the Laurentian Ice Sheet. Over time, climate warming irreversibly melted the southern flank of this compressed mass of frozen water.

The bare ground on the edges of the Laurentian Ice Sheet mainly consisted of sand and stones of all sizes. This gigantic mass of ice generated its own climate with the temperature difference between the layer of air above and the surrounding air generating lots of wind.

When the wind rubbed against the rough, uneven surface of the ice, it whistled and sometimes roared, as it rushed raging, storming and thundering, into the glacier's deepest crevices.

Here and there, at the base of the glacier, lakes of varying depths formed, some of which were connected to the Champlain Sea. There was initially no discernible life in these lakes: no invertebrates in the cold meltwater, no plants on the banks, and no birds in the sky. That meant no food for herbivores and no prey for carnivores. As a result, there were no sounds from living creatures in these sterile surroundings—for the time being, at least.

When the winds eventually subsided, we can imagine the *pluk-pluk-pluk* dripping of the melting ice most exposed to the sun's still-feeble rays. Activated by the drips, waves would ripple out, shimmer briefly, and then fade away with a soft lapping sound on the stony shore.

Far beneath the glacier, we can also imagine the muffled groaning of ice subjected to the enormous pressures from above, or the trickle of cold grey meltwater stroking the stones crushed by the movement of the colossal ice mass.

As the Laurentian Ice Sheet retreated northwards like a giant planer, it crushed everything in its path, leaving behind deposits of all kinds and shaping the geological landscape for centuries to come with a series of lacustrine plains, fjords, eskers, drumlins, and kettles, stretching to the north.

As temperatures rose, giant boulders occasionally broke off and crashed into the water, creating colossal waves. These thunderous rumblings and brutal clamours were all sounds associated with the time of the great climatic warming.

✦

Fast-forward to around 7,000 years ago and to somewhere on the Canadian Shield's exposed bedrock.

Over the centuries, the increasingly diverse vegetation to the south of the glacier produced millions of seeds every spring. These were often carried by winds or birds' feet or plumage to freshly exposed land far to the north.

As the glacier receded, the climate became milder and the winds gentler. Slowly, larch and black spruce, accompanied by a cohort of shrubs like glandular birch, bog Labrador tea, sheep laurel, serviceberry, and American mountain ash took over the land.

The sound of the winds also changed, becoming more temperate and not roaring as often as before. Instead, winds murmured among the low branches of conifers or whispered as they stirred the leaves of poplars and birches. As the winds became louder, they hissed and bounced off tree trunks.

Animals began to return to the land. And all that wildlife made lots of noise: flying insects hummed, caribou and moose bellowed, birds rustled their wings, and grey treefrogs croaked in chorus every spring.

Not to be outdone, water bodies resonated with the *whauk-whauk* sound of snow geese, the *honk-honk* nasal yapping of Canada geese, and the low *quack-quack* of ducks.

Wet groves vibrated to the monotonous, metallic trills of swamp sparrows and the cheerful *wistiti-wistiti-wistiti* cries of common yellowthroats.

For their part, forest depths echoed with the yelping of foxes and the growling of Canada lynx.

✦

Fast-forward to 2,300 years ago and the base of a steep cliff on the edge of a lake, somewhere on Quebec's Upper North Shore between Tadoussac and Pointe-aux-Outardes.

The slow, powerful and haunting voice of the shaman accompanies a small group of nomadic hunters to the sound of a ceremonial drum.

During the group's seasonal travels many moons before, the *kamanitushit*[2] had come across this place with its unique and supernatural acoustic qualities. The echoing and reverberating qualities of the site, in the hollow of a narrow channel linking the northern and southern sections of a vast body of water, greatly enhanced the ceremony's mystical character. The place itself was conducive to the shaman's easily entering into a trance to communicate with the spirits, who would

[2] "Shaman" in Innu-aimun

then let the group know where game was plentiful with lots of moose and American black bears, as well as the right times to begin the seasonal migration to the hunting grounds.

For several more millennia, humans hunted, fished, and gathered in that part of ice-free Upper America in search of caribou, beaver, arctic char, blueberries, serviceberries, and black chokeberries.

In one documented case, people of the Meadowood culture set up a temporary winter camp not far from a cliff face. They made good use of tools like square-based spear points or triangular hide scrapers fashioned from Mistassini quartzite or Onondaga chert. These discovered artifacts are concrete proof of an extensive trading network with other groups to the south, north and west of this immense territory. Entire sections of this cliff face are decorated with red ochre motifs consisting of over a hundred anthropomorphic or zoomorphic pictograms painted by the group the previous winter. The motifs feature a variety of representations: figures with outstretched or crossed arms, holding various objects; other figures seated in a long canoe; an otter-like animal; a fish with prominent fins; and even a snake. Other more bizarre motifs include a supernatural figure with horns or large decorated ears, arms raised to the sky. A *kamanitushit* perhaps?

✦

Fast-forward to 1543 and to the Strait of Belle Isle at the mouth of the Gulf of St. Lawrence where a whale, after being spotted by some Basque fishermen, is being chased by a harpooner poised in the prow of a small rowboat.

There are the familiar sounds of the hunt: the frantic beast's breathing each time it comes up for air, the sailors' cries, the splashing of oars vigorously dipped in and out of the cold water, the whistle of the unleashed harpoon before it pierces the large mammal's flesh, and the roiling waves as the whale tries to flee, dragging along the boat still tied to it by the harpoon's sturdy line.

At some distance, a little farther out on the waves, the whaling ship awaits the outcome of the mad race: gulls cackle, wind slaps the ship's sails, and its pulleys, wooden frame, and both masts creak.

The exhausted whale is finally dragged ashore to be cut into pieces. Its blubber and flesh are loaded onto the ship where it will be sold back in Europe to make oil, soap, and other products.

✦

Fast-forward to 1802 and to somewhere on a wooded knoll not far from the Ottawa River in western Lower Canada.

With a horrific crash of branches and trunks, another huge Northern white pine has just fallen in combat. For weeks now, the forest from sunrise to sundown has been immersed in the incessant din a continual and frenetic onslaught of axes and crosscut saws harvesting the finest timber.

Logging has become one of the colony's largest economic sectors. The tallest pines are being felled, transported along waterways to the port of Quebec, and then shipped to England. The Royal Navy's appetite for tall ship masts was insatiable at the time.

✦

Fast-forward to 1931 and to somewhere in Abitibi.

A deafening explosion shakes the earth. A new section of a gold mine operating at full capacity is being cleared because this precious metal is considered an ultra-safe and high-demand investment in both American and European markets struggling with a painful economic crisis.

Unlike other collapsing economic sectors at the time, the mining industry is booming. *Drill, baby, drill!*

The caribou bound away from these cursed woods in giant leaps as fast as they can.

The caribou keep a low profile as their broad hairy hooves quietly brush, as if with a sigh, the sphagnum moss of the peat bogs or the fresh snow on the boreal forest floor.

✦

Fast-forward to 2024 and to somewhere on the Island of Montreal.

A perpetual urban soundtrack in the background—a veritable cacophony of vehicles honking their horns and revving or slowing down their engines, combined with the puff-puff of compressed-air brakes, the staccato of jackhammers, the wailing of emergency sirens, and the blast of overly loud sound systems in vehicles vibrating under the onslaught of low-frequency noise.

City traffic never stops, and the city's heart beats 24/7.

You have to be very much on your toes if you want to hear a song sparrow ensconced in urban park shrubs as it chirps its heart out all day long in a vain attempt to attract a female to its territory.

Even the pigeons, in the middle of the city, are forced to coo as loudly as possible to be heard.

Long gone are the days in Upper America when the wind whispering through the spruce trees was the loudest sound to be heard.

2

Soundscapes of the Four Seasons

Is the forest totally silent in winter? *By no means!* Yet, due to the slower vibrations of the molecules that physically support sound waves, sound does travel more slowly through cold winter air and also travels farther.

Instead of getting lost in space, sound waves move more easily in winter because they bounce off the frozen surfaces of rocks, tree trunks and bare hillsides. Leafless trees and shrubs, as well as snow-covered ferns and herbaceous plants also have the same effect.

When it's very cold in the forest, and animals fall silent, wind creates soundscapes from tall, bare, dormant deciduous trees.

Intertwined maple and beech boughs groan and creak as they rub against each other, while the bent tops of small trees like cherry, striped maple, or American hornbeam halfway up the canopy often rasp against the trunks of oaks or basswoods, producing screeches reminiscent of chalk on a blackboard. Even the slightest breeze makes conifer needles sigh and there is always the eternal rustling of tall pines and aging spruce. And when sleet falls, the whole forest awakens with a damp spluttering.

Due to frost cracks in tree bark caused by extreme cold, tree trunks sometimes suddenly splinter with a crack like a gunshot. More often than not, these are fungi-marauded trees, whose waterlogged and partially rotted insides have expanded under the effect of frost to burst their outer bark.

There are also times when the forest manifests a silent, serene calm, especially when there's no wind and any sound is absorbed by layers of snow on the ground or by snowflakes falling from the sky.

At such times, we sometimes hear from afar the *tock-tock-tock, tick-tick-tick, tock-tock-tock* sound of woodpeckers drumming on tree trunks. Is this some kind of secret Morse code that lets woodpeckers communicate with each other over long distances? Not at all, because when woodpeckers drum with their beaks during the cold season, they aren't performing any communication function at all—they're simply listening to the timbre of the sounds produced to determine a particular tree's state of health. A sick, totally rotted tree makes a different, less dry sound than a healthy tree does. For hungry woodpeckers, a diseased tree comes with the promise of a feast of insects.

What's that furtive cackling at the edge of a wood, followed by a loud, rolling, discordant sound, reminiscent of a rusty clothesline pulley? A solitary blue jay? No, it's actually several because the species becomes gregarious in winter, even patrolling their habitat in clans of four to six individuals in mid-season.

Jay-jay, says one.

Tree-ooloo, tree-ooloo, responds the other.

Although blue jays have a varied repertoire of cries, they are matchless mimics that can at times perfectly imitate the calls of red-shouldered or red-tailed hawks.

When a small blue jay flock comes across a sparrowhawk or an owl, their excitement is palpable. They ramp up their calls of alarm to warn their fellow clans nearby. They can also perch close to a sparrowhawk, which normally remains motionless, often at the top of a tree or concealed behind a conifer trunk. The jays then continue their cries or swoop down on the predator, feigning a direct attack, although they rarely follow through. Most sparrowhawks eventually get fed up and head elsewhere.

In another part of the woods, a faint *chickcadee-dee-dee* is heard from some undergrowth backlit by the sun. It's the contact call of a small flock of black-capped chickadees on the move. Like blue jays, chickadees are sociable at this time of year.

The standard pattern is for small flocks of six to ten chickadees (consisting of a breeding pair, their offspring from the previous summer, and a few unpaired adult males or females) to criss-cross a winter feeding territory covering about eight hectares[3] (0.08 km^2). Each flock defends its borders and food sources against any intrusion from neighbouring flocks. Chickadees also use a wide range of sounds to conduct their interactions with each other, maintain dominance hierarchies within clans, and dictate behaviour in the event of danger or threat.

Chickcadee-dee-dee, says a chickadee perched in a young, dense fir grove, momentarily separated from her fellow chickadees and trying to gather them around her.

Tseet-tseet-tseet, say a few others, as they quietly feed on grubs hidden in dried goldenrods at the edge of a snowy wood.

Dee-dee-dee-dee, cries another chickadee excitedly, as it drives a rival from another flock away.

One or two downy woodpeckers or white-breasted nuthatches often join these small chickadee flocks to form mixed groups that travel slowly through the forest looking for food sources and other opportunities. Why do such multi-species groups form at all? One reason is that the more eyes from each species, the greater the likelihood of detecting birds of prey, like hawks or crows, and then sounding the alarm more promptly.

Year-round, agile and quick white-breasted nuthatches are also great chatterboxes, as they fly down headfirst from bare tree trunks in search of insects nestling in tree bark folds. In January, they can be heard uttering their powerful, nasal cackle: *hank*, *hank*, *hank-hank-*

[3] Approximately twice the size of the normal summer breeding territory area of four hectares per breeding couple on its own

hank-hank-hank-hank! To the human ear, nuthatches can sound like little clowns hidden in the trees, laughing at us clumsy bipeds in the snow below.

A time when sunlight becomes stronger and warmer, February marks the start of mating season for nuthatches: the male courts the female by bringing her a morsel of food as a nuptial gift, while making a rising and falling whistle sound just before chasing her among the branches in a prelude to mating.

In early spring, as the deciduous forest, with its avian inhabitants in their blue, black, and white feathers gradually becomes a sea of conifers from south to north, the birds' vocal register changes and their feathers turn into a plethora of greys, browns, and reds as other bird species become more dominant in more northerly areas: grey jays (or Canada jays) as opposed to blue jays, boreal chickadees as opposed to black-capped ones, and red-breasted nuthatches as opposed to white-breasted ones.

These three species have chosen a different, colder, harsher habitat than their southern cousins. Still, their genetic affiliation is reflected in a vocal repertoire comparable in almost every respect.

Slightly less variable than those of its blue jay cousin, the grey jay's clucking, soft whistling, and raucous cries still reflect the vocal plasticity of the corvid family[4] to which it belongs.

Like their southern cousins, boreal chickadees also make *chicka-dee-dee-dee* sounds, but these are slower and more nasal. In fact, they are more of a weak *tsick-tsick-dee-dee*, as if the bird lacked confidence. The same goes for the red-breasted nuthatch, whose nasal cackle, *heink, heink, heink-heink-heink-heink-heink-heink!* is looser and more laidback than that of its white-breasted cousin. It's as if the two species try to avoid spending too much energy squawking at each other, all the better to survive in the hostile and biologically unproductive environment of the boreal forest, even though it's the largest land biome on the planet.

[4] Magpies, jays, crows, ravens, etc.

The boreal forest ecoregion covers vast areas of North America, Europe and Asia, and can be considered a world in itself. In Canada alone, the boreal forest stretches over 3.2 million km^2 from the Yukon's northwestern border to the eastern tip of Newfoundland.

The boreal forest in Quebec, known as the Boréalie, is a diverse expanse of lakes, rivers, wetlands, open and closed forests, and dry and arid barrens. Covering a little more than 1 million km^2,[5] from Hudson Bay to the Labrador Sea, and spanning 10 degrees of latitude between the 48th and 58th parallels, the Boréalie forms the geographical heart of Quebec (70.6% of its land area).

Despite being the largest wilderness area of northeastern North America, over 90% of Quebecers have never set foot there.

The Boréalie is an immense territory, a forested area larger than that of Sweden, Finland, and Norway combined. This ancient and mysterious biome is populated by only a few forest tree species: ubiquitous black spruce, as well as jack pine, balsam fir, white spruce, tamarack, white birch, balsam poplar, and trembling aspen. It is a cold forest where willow, serviceberry, and other ericaceous shrubs, such as blueberry, bog Labrador tea, and sheep laurel, grow slowly on thin, poor, acidic soils dominated by sphagnum and other moss species.

Organic matter decomposes slowly on the boreal forest floor due to the scarcity of decomposer organisms (invertebrates, microorganisms, etc.).

The animal species that live there have adapted to the forest's harsh climate and its marked temperature differences ranging from freezing cold in midwinter to stifling hot in midsummer. These species also must cope with major disruptions like forest fires or insect infestations, which often give the forests a uniform appearance. This is because trees start growing again at the same time and are thus of the same age after such disruptions.

[5] 1,068,400 km^2, to be precise

The Boréalie is home to wolf, marten, and Canada lynx, as well as the hundred-year-old beard lichen that clings to the branches of equally venerable black spruce.

Canada lynxes are known as the "ghost of the boreal forest" because they move noiselessly over snow and moss. The underside of their paws is very hairy, and the fur between their pads and toes acts like a snowshoe, not only giving them good balance on fresh snow, but also making virtually no sound. In this way, lynx can surprise their prey, including even snowshoe hares, which, despite their acute hearing, don't always hear lynx coming. Intrinsically solitary creatures, lynx don't vocalize much, except during the breeding season, when both males and females howl and moan.

The great-horned owl is another discreet bird of prey that haunts the boreal forest. Usually operating at night with its unique wings muffling the sound of their beats, this owl hunts hares in silence.

The great-horned owl's low, melancholy *hoot-hoot-hoot-hoot, hoot-hoot,* can be heard in the early hours of clear nights. With this sound that is both a territorial marker and a love song, male great-horned owls can gently approach the female that is the object of their intentions. If she's not too aggressive and tolerates his presence, he moves on to the next stage of closer courtship during which a variety of vocalizations from either partner is accompanied by gentle mutual pecks on the edge of the mate's mandibles or elsewhere on their head. Then comes the mating.

Great-horned owls don't build their own nests; instead, they make do with the abandoned nests of crows, ravens, hawks, or great blue herons. Starting in late February or early March in the Laurentian foothills, and in April in the Boréalie, the males clear away snow to let the females lay her eggs, which hatch a month later. Newborn chicks can often be heard, incessantly wailing and crying out to their parents for food.

Since snowshoe hare populations regularly decline every ten years and snowshoe hares are the great-horned owl's main prey, the great-horned owl's survival and reproduction rates drop in tandem at those times; the bird even stops breeding.

The boreal forest tends to thin out as it stretches north: conifers diminish in size and the ground is covered with an unbroken greenish-white carpet of lichen as far as the eye can see.

This is the taiga ecoregion, home of the enigmatic wolverine, an animal that looks like a mean small bear—the boreal version of the Tasmanian devil. Mainly scavengers during winter, wolverines feed on the remains of dead animals, including those of caribou or moose killed by wolves. On occasion, they are even bold enough to steal a fresh kill or confront an entire wolf pack. Few people have heard the wolverine's vocalizations, but it is said to growl, snarl, and bark like an angry pit bull.

Farther north lies the tundra, an ecoregion where trees become fewer and fewer, and the climate increasingly brutal. The tundra is very dry, with only 400–500 mm of annual precipitation at this latitude. Only three conifer species grow here: larch, black and white spruce, and the lone hardwood, balsam poplar. This region is the Eden of a bygone interglacial era that witnessed the great transhumances or seasonal migrations of the caribou, the sole surviving herbivore of yesteryear's American megafauna, which included the woolly mammoth, the Ellesmere camel, and the American mastodon.

Tundra caribou are highly gregarious animals. During the long Arctic winter, they dig into the snow crust with their hooves to feed on the lichen. They grunt and snarl as they compete with each other for the best feeding sites. In fact, the larger males sometimes force away females, young herd mates, and the year's calves, with the latter not hesitating to bleat or bawl to their mothers for help.

This tundra is a soundscape as yet undisturbed by humans—for now.

✦

As for southern Quebec, when spring finally arrives along the St. Lawrence River, the sounds of the forest change completely, starting with the wind that no longer whispers through the conifer needles, but rustles leaves as if with a lisp. Indeed, the entire natural soundscape changes as a new procession of mammals, birds and insects enters the scene.

On the flora front, the terrain's higher slopes are dominated by sturdy northern red oaks and proud northern white pines. The mid-level slopes are home to maple forests that contain a variety of trees—not only sugar maple but also several other species like basswood, American beech, and yellow birch—that reach 30 m or taller. On the lower slopes, the predominant tree species are Eastern hemlock on deep, cool soils; black ash on swamps; and cedar bushes on more acidic peaty soils.

With the arrival of spring, water, freed from the cold and once again in liquid form, feeds new streams and torrents in the undergrowth. The young, regenerated mixed forests that rise on the banks of slow-moving rivers and in the depths of quiet bays become the uncontested domain of the Canada warbler, a small, sparrow-sized forest passerine with two tones, bluish-grey on the back and yellow on the throat and breast. This charming bird, sporting, as it were, a gleaming necklace of black sapphires (like Audrey Hepburn in *My Fair Lady*) sings a rapid staccato with irregular phrasing that begins with a *chip-chip!* sound, somewhat evocative of a brook's sparkling cascade.

The rest of the fauna soundscape is composed of a wide range of sounds: the grumbling of post-hibernation black bears, the whistling of marmots, the cawing of crows, and the croaking of toads. There is also the dawn chorus of whistling robins, cackling grouse, cheeping sparrowhawks, gobbling turkeys, squeaking voles, yelping foxes, babbling wrens, chirping vireos, twittering chickadees, howling coyotes, growling bobcats, and believe it or not, humming mosquitoes, and buzzing black flies!

As night falls, it's the evening choir's turn to perform its repertoire: a hermit thrush on a lakeshore chants its ethereal song, beginning with a long, fluttery note and invariably ending with increasingly higher tirades; spring peepers cry out at the tops of their voices in the cold water of springtime ponds; and opportunistic salamanders, guided to the pond by grey treefrogs' croaking, slip silently in for their only lovemaking night of the season.

In southwestern Quebec, the species-richer terrestrial ecosystems are dominated by sugar maple forests. These forests are divided into three distinct types from the 45th parallel to the 47th: bitternut hickory-maple, basswood-maple, and yellow birch-maple.

Stretching like a tiny triangle across the extreme southwest of Quebec, a swath of bitternut hickory-maple forests make up the richest and most diverse forest area in all of Quebec. Its springtime flora, which explodes in the undergrowth in May, includes white trillium, two-leaved mitrewort, bloodroot, sharp-lobed hepatica, two-leaved toothwort, and large-flowered bellwort. However, as soon as sugar maple, oak, and hickory foliage spreads over them and deprives the forest floor of sunlight, these ephemeral beauties disappear.

The swath of bitternut hickory-maple forests, lying on top of soils rich in sediment, clay deposits, and littoral sands inherited from the Champlain Sea, are home to 46 native tree species and a vascular flora of some 1,600 species. Many of these trees, like white oak, black maple, hackberry, bitternut hickory, and shagbark hickory, are at the northern limit of their range.

Sugar maple is still the dominant tree species in the basswood-maple forests that cover much of the St. Lawrence River lowlands as far as Île d'Orléans, the Eastern Townships, and Bois-Francs, accompanied by other forests of basswood, beech, ironwood, white ash, or yellow birch, depending on soil conditions. Fewer tree species (41) and vascular plants (around 1,500) grow in basswood-maple forests.

Tree diversity in yellow birch-maple forests drops to 23 species, and total vascular flora to an estimated 900, due to these forests' rugged topography and higher average latitude. Wherever the soil is neither too dry nor too wet, the mature forest is dominated by sugar maple, followed by yellow birch, with beech, fir, red spruce, and white spruce also part of the mix. In the northwestern corner of Témiscamingue's yellow birch-maple forest, there are virtually no beech, whereas red maple is abundant to the east. A good-sized shrub (striped maple) and a herbaceous plant (cucumber root) are also ubiquitous.

These differences in the plant communities of temperate Quebec naturally have repercussions on the animal communities and, in turn, on their soundscapes.

While the songbird choirs in bitternut hickory-maple forests include great crested flycatchers, wood thrushes, brown thrashers, and eastern towhees, those in their yellow birch-maple counterparts include blue-headed vireos, olive-backed thrushes, black-throated blue warblers, northern parulas, and Nashville warblers. In other words, the hour of dawn sounds very different, depending on latitude.

Whip! is the loud, strong whistling sound from a great crested flycatcher that resonates high in the tree canopy of a mature wood south of Montreal. The sound is heard twice before being followed by a long pause.

Bleu-bleu-bleu gree-ees! is how a black-throated blue warbler sings in the depths of a Laurentian forest in the Canadian Shield.

✦

It's difficult to talk about temperate forest soundscapes without recounting what happens beneath the ground.

A healthy soil containing a wide range of fauna, fungi, and flora provides basic ecosystem functions, such as the decomposition of plant materials and the cycling of nutrients. The greater the diversity of living organisms, the more resilient the ecosystem and the faster it can return to its initial state of integrity after a disturbance, even if individual species are missing, since their role in the soil can be played by other organisms. In southern Quebec's fertile soils, there are so many plant and animal species that, with the right equipment, one can hear the sound of microfauna literally moving and swarming beneath one's feet. Countless arthropods, mites, beetles, ants, centipedes, millipedes, and springtails (earth lice) grind, crush, slice, and decompose organic matter from dead leaves that fall to the ground each fall and pursue the organisms that feed on them. For example, tiny roundworms called nematodes feed on decomposing waste and even creatures smaller than them (like rotifers); and larger animals—moles and shrews—go after spiders, insect larvae, and other tasty creatures. This soil fauna makes a lot of noise as it moves about among the fallen leaves or beneath the ground.

The soil itself is also used by different species to communicate with each other. These underground inhabitants find ways to generate vibrations that their fellow creatures since pick up via legs or other body parts (such as cilia, antennae, and vibrissae).

These days, biologists, acousticians, and other researchers use specialized microphones to record and statistically estimate the proportion of sounds in the soil coming from living organisms as opposed to sounds from other sources, such as precipitation absorption, wind impact on surface vegetation, or man-made noise pollution (e.g., vibrations caused by proximity to roads or airports). Even the deep roar of jumbo jets soaring through the sky can drown out the noises made by microfauna in the soil.

However, we still know very little about the effects of noise pollution on the distribution, activity, and composition of soil microbes.

✦

Our first scenes are set in the Laurentian foothills. We start on a lakeshore at dusk on a mid-July day. *Plook!* is the sound a brook trout makes as it jumps out of the water to capture an insect and then slips back into the water.

In a shallow bay of the same lake, near a stream meandering through dense undergrowth dominated by speckled alder, a wood turtle rhythmically taps the ground with its front paws to imitate the sound of raindrops falling on the forest floor. This is a ploy by the patient reptile to make earthworms believe that it really is a downpour; they can be gobbled up when they come to the surface.

Not far from the turtle, an American woodcock makes its characteristic metallic *pinzt!* sound to indicate its presence in the shade of a thicket, while, higher up in the branches, a flaming warbler emits its *zooee-zooee-zooee* motonone.

At the same time, near a mud-and-branch dam built across a stream, a solitary beaver, startled by a suspicious noise on the bank, uses its tail to make a sudden, muffled *clap!* on the water surface to warn its little family of danger—maybe a wolf, bear, or other predator is lurking. If the threat is real, and a predator corners the beaver in its territory, the beaver will belligerently growl or snap its teeth.

Sound moves faster through water than through air thanks to the acoustic energy of the vibratory waves traversing the ambient medium. The speed of sound propagation in water increases as a function of three factors specific to water: temperature, salinity (e.g., brackish or salty), and acidity (especially in the case of low-pitched sounds).[6]

[6] Rising ocean acidity is a growing concern for the scientific community. The culprit? Global warming, exacerbated by CO2 emissions that react with water to form carbonic acid. Since the increased presence of ions in water (due to its rising acidity) supports and accelerates the movement of low-frequency waves over long distances, there is concern that anthropogenic sounds (for example, noises associated with deep-sea oil operations, or with the engines of large ships) may interrupt cetaceans' ability to communicate and move

Ambient pH significantly impacts freshwater organism communities, given that plants and animals often have fairly strict requirements for alkalinity (high pH values) or acidity (low pH values). Most lakes have values ranging from six to nine on the pH scale, although the water in some areas surrounded by peat bogs is acidic, with a pH value of four or less. In the St. Lawrence River lowlands, and in most of the Appalachians, streams flow over clay, limestone, or shale, which makes the water more alkaline. However, the pH of streams flowing over granite rock throughout the Canadian Shield (i.e., the lands starting with the foothills north of the St. Lawrence River) and in a few places in the Appalachians is neutral under natural conditions.

Peat bogs, which are acidic environments par excellence, host a unique collection of flora and fauna. In spring, they resonate with the rusty blackbird's grating *chacks!*, reminiscent of a poorly-oiled hinge, or the rufous-crowned warbler's buzzing, high-pitched trill.

In a wide river's slow-moving brown waters in the St. Lawrence Valley to the south, carp begin their mating rituals. In the full swing of spawning season, males make lots of noise as they splash on the surface, fighting with each other to inseminate the eggs that the females have randomly released.

Close by, a great blue heron impassively lies in wait for the gastronomic opportunity offered by just one of these large, carefree carp.

Farther afield, a group of northern river otters frolic, chuckle and snort among themselves in their typically social fashion before setting off to hunt for bullheads, sunfish, suckers, redhorses, and other bottom-feeding fish. Working together, the otters can corral and compress schools of small fish into shallow bays where all exits are blocked. Unlike solitary feeders, otters catch more fish by hunting in groups; in

---

about, and also harm certain fish that orient themselves in their coral reefs by listening to the sound of sea waves against coral. Sound travels further in water in response to a pH increase of just 0.3. According to the International Fund for Animal Welfare, noise measurements taken in the northeast Pacific indicate that anthropogenic noise has increased by 200% since 1950 in areas where maritime traffic is most intense.

this way, they also often add more variety to their diet by catching faster-moving species. Otters communicate over long distances by making high-pitched chirps and, when disturbed by an intruder, initially respond by barking like a puppy and, if the threat is more serious, express their anxiety and nervousness by sniffing loudly.

As the day draws to a close, a male carp, exhausted by his breeding exertions and oblivious of the statuesque great heron standing nearby, rests near the surface, close to the shore.

Splash! In a single lightning-quick swoop, the great blue heron captures the carp in its slender beak before swallowing its prey in one gulp. The carp's vainly struggling body continues to be visible as it moves down the narrow corridor of the wader's long neck.

Now comfortably full, the bird watches the sun go down behind the trees lining the river. It will soon perch for the night to digest its meal.

An extensive watershed consisting of streams, rivers, torrents, lakes, ponds, marshes, swamps, and peat bogs follows the topography of the land between the mountains and the sea. All these wetlands are unique in terms of not only their water properties but also the variety of living species that occupy these aquatic habitats.

✦

Our next scenes take place in late September on the high hills of Matane's hinterland, where the Appalachian Mountains and the cold, humid fir forests of the northeastern part of the continent culminate in the Gaspé Peninsula.

High in the sky, a burly silhouette hovers in wide circles above the foothills: a small broad-winged hawk is on its way south. Every now and then, for no apparent reason, it lets out a long, high-pitched whistle, *whee-ee-ee-ee!*, and then disappears over the horizon.

A family of ravens—the female, the male and their latest offspring born earlier in the year—are patrolling a ridge covered with only a few scattered black spruce trees. In a brief series of low, raucous caws, they exchange useful information with each other.

For the region's moose, the rutting season is in full swing, with the cow moose having her first ovulation between the third week of September and the first week of October. With a brief 7–12-day fertility period, she has no time to lose and scours the vicinity to attract a bull moose by urinating and uttering long, nasal moans.[7] Some distance away, in the fresh windless air of dawn or dusk, a bull hears the sound of the cow splashing in a marsh, or urinating into water or onto the ground. He scrapes his antlers against tree trunks to attract the cow's attention and warn off rival males in the vicinity. When a bull and a cow finally meet at the right moment, the male approaches slowly and emits brief hiccups, produced by rapidly opening his mouth to suck in air and then smacking his lips. Just before mating or when the male is close by, the female moos and moans softly to soothe him.

As the days get shorter, the American black bear is insatiably hungry because he knows that winter is fast approaching. It's time to build up his fat reserves so that he can make it through to spring. Any food source—wild serviceberries, beaked hazelnuts, beechnuts, insect larvae, or even a little carrion—fits the bill. If the bear is disturbed by a human or other intruder during its feeding, it grunts or snaps its teeth. Then, if the intruder doesn't get the message, the bear feints in the intruder's direction or loudly taps with its front paw on the ground or on another object. This is usually enough to drive the intruder away. As the calen-

[7] The call of a cow moose in heat lasts 0.6–7.5 seconds, depending on her hormonal state and the presence or absence of a bull moose in the vicinity. Echoed by rocky cliffs, the cow's call can be heard in a radius of several kilometres, rushing through valleys and bouncing from tree to tree.

dar moves deeper into October, the bear gets clumsier and clumsier, eats less and less, and looks for a den under a large tree or in a cavity between rocks where he can hibernate for the winter.

In the shadow of the fir forest, under the withered ferns of fall, two southern red-backed voles—a dominant adult and a younger one—meet by chance on the forest's mossy floor. In a brief exchange, the former threatens the latter with a high-pitched scream. Accepting this domination, the latter squeals and anxiously moves away, chattering its teeth.

In a nearby natural cave halfway down a rocky scree slope, some little brown bats and a few northern long-eared myotis share the hibernation site they've been using for the past several years. Although the colony is noisy, we can't hear a thing because the sounds of the bats gathered together are beyond the range of human hearing.

Other settings, other customs: in a vacant lot in a bustling metropolis some one thousand kilometres away, some grasshoppers and crickets are making themselves heard above the city's uninterrupted din. In the warm light of the fall equinox when both the day and the night are twelve hours long, they are singing their last love songs of the year.

Second Movement

The Nature of Sound

... Life is noisy, and only death is silent...

Jacques Attali
Bruits

3

Geophony, Biophony, and Anthropophony

No matter where you live on Earth, the soundscape that envelops you is the combination of three distinct sources: natural phenomena (geophony), living organisms (biophony), and human beings (anthropophony).

Geophony comprises an infinite range of natural sounds: thunder, rain, sleet, waterfalls, roaring rivers, tall pines trembling in a sudden gust of wind, reeds swaying lazily to a breeze, and waves crashing on the shore. Geophony is the sound of sky, earth, and sea combined—an ancient voice, more than 500 million years old—a voice that was alone on Earth for a very long time. A surprisingly varied voice as well. Take rain, for example. Depending on how much there is, rain can be unobtrusive, calm and ethereal, but it can also be brutal, deafening, and threatening. When rain falls on large rocks or a stony shore, it makes a hard, metallic sound whereas when it falls on the broad leaves of a basswood tree or a water lily, it makes a soft, organic sound. Watercourses like streams and rivers also produce a rich variety of sounds, depending on how fast the water is moving, the nature of the stream bed or riverbed, and the types of rocks or pebbles the watercourses contain.

Biophony is the sound of life, the totality of sounds made by living organisms, big or small. It consists of the sounds made by the auditory organs of a wide range of species that have emerged over millennia through natural selection and the blind hand of evolution. Here are some examples. The grasshopper produces its chirps in summer by rubbing the back of its femur against the edge of its wings in a rapid back-and-forth motion. The male wood thrush uses the muscles of its syrinx, a small resonating chamber at the split in its trachea, to produce fluty, richly modulated sounds to attract a female or mark out its territory during the breeding season. To orient themselves or help them capture their favourite fish, belugas use their jaws and the fatty, fibrous tissues of their bulging foreheads to produce ultrasound waves too high-pitched for the human ear.

Anthropophony is the totality of sounds generated by humans—not only words, songs, and the roar of crowds, but also the sounds of all the tools and processes we've invented since our *Homo sapiens* ancestors first struck a stone to shape spearheads and fashion game meat scrapers.

Since then, there has been an uninterrupted succession of new sounds produced by new tools and activities, such as ploughs, ceremonial drums, steam trains, mining in the nineteenth century, rock blasting in the twentieth, giant combine harvesters, and supersonic planes. Today, this electromechanical backdrop of cars, trucks, chainsaws, lawnmowers, leaf blowers, and factory machinery has become our normal soundscape, along with unpredictable sounds from sources like sirens, bells, rifles, handguns, and even bombs.

✦

The overall increase in decibels from human sources is a direct result of our ever-expanding road, rail and air transport networks criss-crossing the globe, the rapid urbanization of land areas, and the sustained demand for natural resources (such as minerals and forest products).

A growing body of scientific research has documented the impact of man-made noise on animal species in both aquatic and terrestrial environments. The negative effects of human noise on living organisms are varied, affecting many facets of animal life like physiology, reproduction, communication, search for food, and predation dynamics. Man-made noise reduces the ability of many animal species to detect and respond appropriately to sounds in their environment.

Chronic exposure to noise can produce a physiological response in the form of an increase in stress hormones. Scientists have measured this phenomenon in the sage-grouse (a plains bird in the American West): when exposed to high levels of noise associated with gas exploration or road traffic, this species produces 16.7% more cortisol than normal. A 2002 study by five American researchers on the impact of snowmobiles on wolves and elk in various national parks came to similar conclusions. By analyzing glucocorticoid levels in these animals' feces, the researchers found that their endocrine response was linked to exposure to engine noise: the higher the noise level, the higher the glucocorticoid level found in the feces.[8]

Over time, animals subjected to such conditions change their behaviour, as in the case of some male passerines who changed their singing or sang more often in order to continue to attract females until they eventually left their habitat because it had become uninhabitable.

Just as in humans, higher levels of stress hormones like adrenaline and cortisol can reduce animals' physiological resistance to illness by weakening their immune systems. Chronic stress hormone production has an exhausting effect: it upsets other systems at such as those that regulate blood pressure or blood sugar levels.

Noise generated by human activity also has unexpected effects on natural phenomena like plant pollination by insects, the dispersal of tree seeds by birds, and the composition of forest-floor arthropod communities. Honeybees, bumblebees, and other similar insects are less likely to visit flowers in environments where noise pollution is severe.

[8] Scott Creel, Jennifer E. Fox, Amanda Hardy, Jennifer Sands, Bob Garrott and Rolf O. Peterson, "Snowmobile activity and glucocorticoid stress responses in wolves and elk." *Conservation Biology*, 16(3) (2002): 809-814.

Similar impacts have been noted on the populations of certain terrestrial arthropod species (e.g., grasshoppers, crickets, wolf spiders, and jumping spiders) in the USA. When some species are negatively affected, other opportunistic species can take advantage of the situation to proliferate.

Such changes in insect communities, which play important ecosystemic roles (e.g., as pollinators, prey, or nutrient decomposers), can have significant consequences for the health of natural environments. For example, the natural regeneration of forests in the southern United States has slowed because the California scrub jay, a jay species that normally feeds on pine nuts, other hard-shelled seeds, and tree fruits avoids overly noisy environments and no longer helps spread seeds to habitats suitable for establishing new generations.

To establish this link between jays, ambient noise, and forest renewal, three scientists from Texas and California monitored the regeneration of juniper and pinyon pine over a fifteen-year period in a forest stand exposed to high levels of artificial noise in New Mexico. In 2021, they found a 75% reduction in pinyon pine seedlings in noisy areas compared with quiet settings. The scientists then examined areas where artificial noise had been reduced to determine long-term tree growth once quiet had been restored. They found that new growth remained scarce, with jays apparently reluctant to recolonize those areas.

Man-made noise thus has a severe indirect impact on the distribution and numbers of shrubs, trees, and herbaceous plants, which results in cascading ecological effects that diminish biodiversity.

That said, noise pollution from human sources represents opportunities for certain species, either because the areas concerned are abandoned by most predators, or because predators can no longer detect their prey. Species like owls that mainly use their hearing to hunt small mammals are at a particular disadvantage in such situations, to the great relief of mice, voles, and shrews, which can then proliferate unhindered.

Man-made noise profoundly alters the structure and species composition of living communities. According to a Japanese study published in 2016, the sound in decibels (dB) from road traffic reduced the hunting efficiency of owls by 90%. From December 2014 to March 2015, the Japanese researchers observed the predatory activities of 78 owls in the wild (45 short-eared and 33 long-eared—two cosmopolitan species that also nest in Quebec) and analyzed their prey detection rate in relation to ambient noise. The experiment showed that a noise level of 40 dB, the equivalent of a quiet residential area, reduced the owls' prey detection rate by 17%, while a sound level of 80 dB from vehicle traffic reduced prey detection by 89%. According to the researchers, this phenomenon alone can transform the nature of prey/predator interactions, and may negatively impact the ecosystem as a whole.

✦

Biologists have been studying the sounds made by animal species for the past few decades. As soon as the technology was available,[9] ethologists and other biologists recorded soundtracks in the wild to better understand the role of mating sounds of species like whales, grey treefrogs, and other frogs. They also recorded and deciphered the calls of alarm used by passerines to respond to predators, or the territorial drumming of woodpeckers on dead trees. The field of bioacoustics was born.

More recently, the new discipline of ecoacoustics has emerged. This discipline focuses on the entire soundscape of a given environment, such as a lake shoreline, a seabed, a coral reef, or a primeval boreal forest as opposed to the sounds of a single species. Instead of cap-

[9] Around the middle of the last century

turing just a violinist's solo, the whole orchestra is recorded so that the entire score (biophony, geophony and anthropophony combined) can be decoded.

With these new techniques for capturing sounds in natural environments, we can now automatically record soundscapes at specific times. We know, for example, that dawn, followed by dusk, are the best times to capture the greatest diversity of animal sounds. With these techniques, we can also capture sounds in a programmed sequence—five minutes every hour, 30 minutes every 24 hours, and so on.

Digital tape recorders, discreetly installed in selected locations, can capture the rich sounds of a marsh, a peat bog or an old-growth forest over long periods of time.

Some devices record not only sounds audible to humans, such as birdsong, but also other sounds in ranges inaudible to humans, but which are perceived by other species like bats. Scientists can use software to transform recordings into spectrograms that can show a given species' characteristic signals.

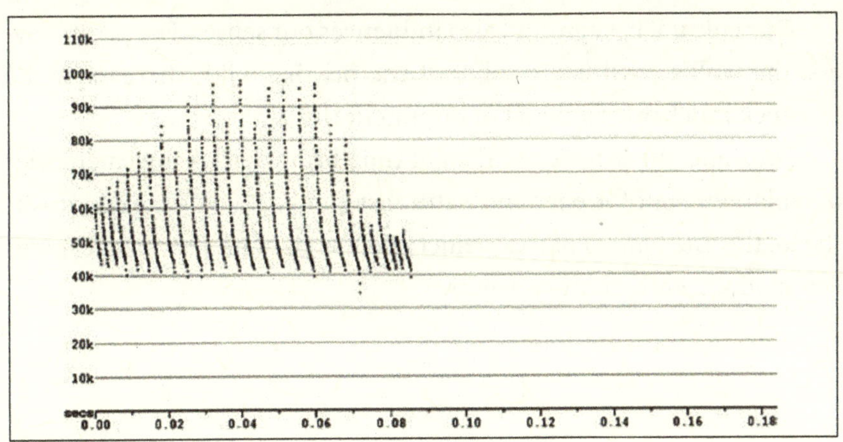

Table 1 A northern long-eared myotis's typical sound signature with each spectrgraphic line corresponding to a bat call. In this case, the Y axis shows the recorded frequency from 40k (40,000 Hz) to 98k (98,000 Hz), and the X axis indicates the duration of each sound emission in hundredths of a second.

The aim of deciphering soundscapes is to obtain valuable information on the ecological integrity of natural environments, and better monitor the volume and distribution of sounds produced by living organisms in order to understand and protect biodiversity.

To do so, it is essential to use the most objective methods and analytical tools possible because our human perception of the sound environment passes through two filters: that of our hearing capacity (humans hear a restricted range of frequencies[10]) and that of our culture.

Like all other living organisms, humans become aware of their environment through various sensory organs (e.g., ear, eye, and tongue) that have been shaped by evolution and natural selection for millennia and which have helped fashion tools suited to our lifestyles. These organs are clearly different from those of other species with the result that we live in a sensory universe distinct from that of birds (who, for example, perceive ultraviolet rays), bats (who perceive ultrasound) or fish (who detect changes in water pressure via the lateral line that runs along their sides).

Our cultural background also influences our senses. For nature lovers, the wolf's howl is a symbol of the pristine wild whereas for the livestock rancher it's a signal of imminent threat.

For some, the whistle of a distant train brings back happy memories of childhood, but for poet and naturalist Henry David Thoreau, it was the loathsome symbol of the ineluctable march of progress and capitalism in nineteenth-century America.

[10] See next chapter.

4

WHAT IS SOUND?

What is sound? Let's turn to physics for the best answer.

According to physics, sound has four defining properties: frequency (pitch), intensity (amplitude), duration, and timbre.

Frequency

Just as a violin string vibrates by being plucked, the vibration of any body, large or small, produces waves. These waves propagate through the surrounding medium, whether air, liquid (e.g., water), or solid (e.g., wood). A sound wave's frequency is how often it oscillates per second. This property of sound is expressed in hertz (Hz), named after the German engineer and physicist Heinrich Hertz, who came up with this unit of measurement in the nineteenth century. Frequency defines the pitch of sound: when the frequency is low (e.g., 50 Hz, or 50 oscillations per second), the sound is low-pitched; when the frequency is high (e.g., 14,000 Hz, or 14,000 oscillations per second), the sound is high-pitched.

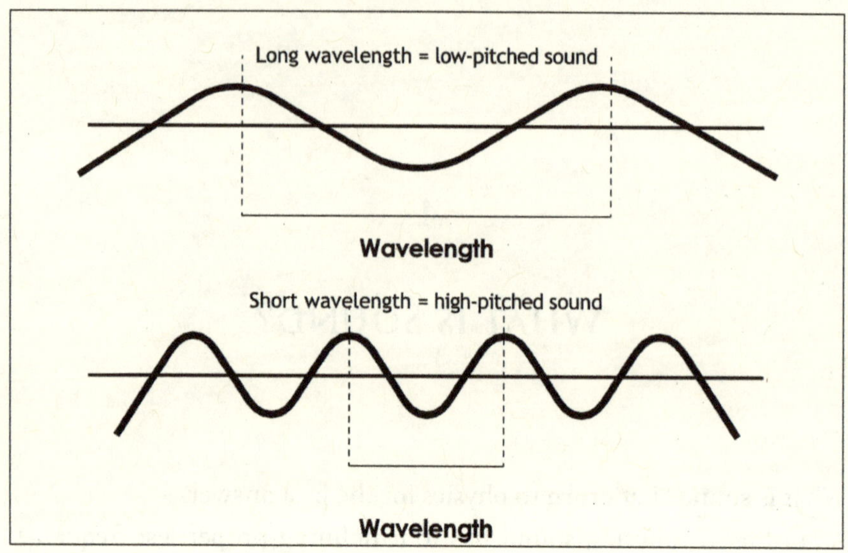

Table 2 A wave's number of oscillations per second, expressed in Hertz (Hz) defines the pitch of the sound: when the frequency is low (i.e. when the distance between two peaks is long), the sound is low-pitched; when the frequency is high (when the distance between two peaks is short), the sound is high-pitched.

A violin string tuned in *A* vibrates at a constant frequency of 440 Hz. This is known as universal *A*, the note used by orchestral musicians the world over for tuning.

On average, the human ear perceives sound frequencies from 20–20,000 Hz, although this varies with an individual's age and heredity. Sounds below 20 Hz are called infrasound; those above 20,000 Hz, ultrasound.

The range of perceptible sounds differs greatly among living organisms, and even within the same animal class. Among mammals, for example, dogs perceive frequencies from 15 –50,000 Hz, and bats, from 15,000–150,000 Hz.

Some species navigate and find food by using a technique called echolocation, which involves producing ultrasounds and then catching the echoes that bounce back from objects. By essentially analyzing echo characteristics, i.e., the difference in frequency between transmitted and returned waves, these species can locate obstacles and other objects, and assess their size and speed (in the case of moving prey). In this way, they can obtain information about their immediate environment and its occupants in a dark cave or ocean depths, for example. This echolocating faculty is shared principally by three distinct mammalian groups: bats (order Chiroptera), toothed whales (suborder Odontocetes), and shrews (family Soricidae).

Bats scan their surroundings for potential prey by constantly generating very rapid pulsatile vibrations (lasting from 5–10 milliseconds) at high frequencies of up to over 100,000 Hz. When prey is detected, the pulsations increase from 10 Hz to 200 Hz per second, and bats continue to receive more information as they get closer to their prey.

Such faculties have also been observed in certain dolphins (white-beaked and short-beaked common), long-finned pilot whales, orcas, belugas, and narwhals. The latter use the same hunting techniques as bats, emitting series of 3–10 clicks per second, rising to 110–500 per second when they spot prey. The northern short-tailed shrew and the cinereous shrew are two other species that uses echolocation—in their case, to find their way in unfamiliar underground environments.

Sixteen species of rare birds also use echolocation. These are mainly in the Apodidae family (swifts and swiftlets), some of which, although diurnal, have been observed fully active at dusk or during the night. However, the chimney swift, the only representative of this family in northeastern North America, does not fall into this category.

Intensity (amplitude)

Intensity or amplitude refers to the intensity of a sound emission. It is measured in decibels (dB), the smallest perceptible unit for the human ear. The higher the number of a sound emission's decibels, the more intense it is.

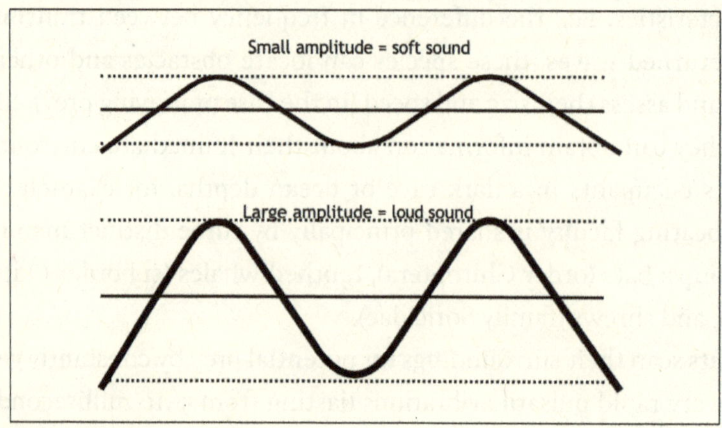

Table 3 The height of a soundwave's peak defines its intensity or amplitude. This intensity is expressed in decibels (dB).

A sound's scale of progression is not linear: its intensity doubles every three decibels. A 15-dB sound is thus twice as loud as a 12-dB sound, an 18-dB sound twice as loud as a 15-dB sound, and so on.

The upper tolerance limit for the human ear is 115–120 dB, equivalent to the noise of a gunshot. Noise levels above 120 dB, can produce tinnitus, temporary or permanent deafness, and other afflictions. Hearing risk in the workplace is estimated to begin at 80 dB if exposure is chronic or long-lasting.

The sound intensity of a single violin at close proximity can reach 92 dB.

Some cetacean species produce very powerful signals which, if emitted in air and not in water, would be equivalent to the discharge of a high-calibre rifle. For example, humpback whales can sing up to 155–189 dB (a vocal range of eight octaves!), narwhals can occasionally click up to 218 dB, and sperm whale can click as high as 223 dB (the most intense sound produced by a living organism). The purpose of these intense sounds is to communicate over long distances between individual animals that are sometimes away from each other in marine environments.

Duration

Duration refers to the measurement, in seconds, minutes or hours, of the period of time between the moment a sound becomes perceptible and when it fades away. However, the sound envelope between these two moments can intensify, weaken, or remain constant. For example, the attack of a violin may be slow before peaking at maximum intensity and then slowly fading away, whereas the attack of a drum, including that of a snare drum, is fast, intense, and quicker to fade away.

The fluty, ethereal songs of both the hermit thrush and the mourning dove begin with a slow crescendo. By comparison, the attack of the sharp-shinned hawk's *keck keck keck keck* is short and dry, and its call ends abruptly.

Some species make the pleasure last: for instance, the mating call of the male walrus, one of the noisiest Arctic animals, lasts from three to seven minutes in North Atlantic populations, which is curiously shorter than that of this species' Pacific populations at two to four minutes. Remarkably, a male walrus can sustain its mating call for up to 48 hours at a time, with little or no break, until the object of its attention succumbs to its charms.

The male winter wren is also a very determined bird with a long, complex 7-10 second vocalization of warbling and high-pitched trills, repeated *ad nauseam* at first light.

Timbre

A sound's timbre brings together the elements (colour, texture, and so on) that give a sound its personality and sonic identity. Two musical instruments can produce the same note, with the same intensity and frequency, but never have the same timbre. For example, it's easy to distinguish a note played on a violin from the same note played on a flute. Timbre is a sound's voice and character, as it were.[II]

[II] In technical terms, a sound's timbre is related to the number and amplitude of harmonic frequencies emitted simultaneously with the main frequency, i.e., physical bodies vibrating

The same goes for sounds from biological sources: each organism that emits sounds has a specific resonance depending on the configuration and type of the organs used to emit them.

How sound travels

As already mentioned, sound needs a physical medium—gas, liquid or solid—through which it can propagate.

Sound waves travel at an average speed of 340 m per second in air but even faster in water and the ground.

Sound fades away in proportion to the distance between source and receiver, as the carrying medium absorbs the sound waves passing through it. The absorption factors for every 100 m a sound travels are 0.008 dB for water, 1.2 dB for air, and 6 dB for ground.

Absorption rates vary according to sound wave frequency. This means that it may be more difficult to hear a distant sound at frequency X than another sound at the same distance at frequency Y. In the particular environment of northeastern America's mature deciduous forests in spring, the song of thrushes at a frequency of 340 Hz travels farther than that of other birds because higher frequencies are absorbed at close range by vegetation and dead leaves on the ground. On the other hand, certain lower frequencies bounce off tree trunks, rocky hillsides and so on.

Acoustic properties vary widely depending on the natural environment of the sound's source. According to a study by a team of Spanish biologists, published in the journal *Animal Biodiversity and Conservation* in 2021, songbirds opportunistically use some of these properties to propagate their mating songs further afield. In fact, in the Mediterranean forest massifs of central Spain, where granite rock formations

---

sympathetically at higher octaves. The second and third harmonic frequencies resonating with an A at 440 Hz are 880 Hz (2 x 440) and 1,320 Hz (3 x 440).

called *pedrizas* reflect sound particularly well, the researchers noticed that several species, including Eurasian chaffinches, Eurasian blue tits, great tits, Eurasian blackcaps, European robins and common blackbirds, sang more frequently near these *pedrizas* than elsewhere in their habitat.

As musicians and naturalists know, low-pitched sounds travel farther than high-pitched ones. For example, a great grey owl's hooting travels for miles in the boreal forest, as does the mournful sound of a rutting cow moose in the fall; and a sperm whale's clicks can travel up to sixty kilometres or more in the sea.

This would explain why passerines like the white-throated sparrow and Swainson's thrush that nest and breed in forests generally emit territorial and mating songs at lower frequencies on average than the songs of birds in open environments or those like field sparrows or golden-crowned kinglets that nest and feed in the higher tree strata where sounds carry further.

During breeding season, some bird species within the same family such as Parulidae cohabit in different canopy strata within the same habitat such as a coniferous forest.

In conifers, for example, yellow-rumped warblers occupy the lower strata, bay-breasted warblers the middle, and Blackburnian warblers the upper. The yellow-rumped warbler's loose, two-tone, descending trill is deeper than the bay-breasted warbler's high-pitched, whistled *seetzi seetzi seetzi seetzi* or the Blackburnian warbler's *tsee tsee tsee tsee tsee ee, dzee dzee*.

It's also important to consider that sound wave propagation is linked to variable physical factors, such as climate. The speed of sound increases with ambient temperature and pressure, while favourable or adverse winds affect the distance at which birds' songs or alarm calls can be heard.

Third Movement

Animal Soundscapes

What's that faint barking in the clouds? Sounds like a dog in the distance. Strange, isn't it, how the world cocks an ear at the sound, and wonders what it is? The barking gets louder—it's geese! We can't see them, but they're getting closer.

Aldo Leopold
Almanach d'un comté des sables

5

How Animals Hear

The hearing of most animals is linked to their sense of balance. In mammals, for example, hearing and balance are associated with the same sensory and mechanoreceptors in the ear.[12] Over time, the vertebrate ear has evolved into an organ that provides both hearing and stability, at least for terrestrial mammals, which are more affected by gravity than their marine counterparts.

When sound waves in the form of moving air enter a terrestrial mammal's outer ear, it causes the eardrum, a stretched membrane at the bottom of the ear, to vibrate. This vibration is then transferred to the middle ear's three ossicles—malleus (hammer), incus (anvil), and stapes (stirrup). In turn, the ossicles transmit oscillations to the cochlea, a snail-shaped bony structure in the inner ear. A coiled organ (the organ of Corti) inside the cochlea houses sensory hair cells consisting of bundles of thousands of tiny, rod-shaped cilia,[13] only visible when magnified 10,000 times. When cilia are stimulated by pressure waves generated by vibrations in their surroundings, they produce nerve impulses that the auditory nerve carries to an area of the vertebrate brain called the *torus semicircularis*, where the sound's parameters, including its pitch and intensity, are decoded.

[12] Neurons sensitive to mechanical stimulation

[13] Each human ear normally contains at least 24,000 cilia

Over evolutionary time, the pinnae or auricles (the scientific terms for the visible or outer part of the ears of humans and other primates) have lost their flexibility. However, this is not the case for many other mammals like hare, deer or Canada lynx that have retained the ability to turn their outer ears towards the source of a sound or noise and correctly identify it.

Although birds have no external pinnae, they do have cochleae, which play the same role as pinnae in mammals. While birds can generally hear sounds in the same frequency range as humans (20–20,000 Hz[14]), their hearing is sufficiently fine for them to distinguish subtle differences in sound wave intensity and pitch. This is hardly surprising, given the importance birds attach to songs, calls and other sounds when defending nesting territories and finding mates.

Owls are the avian species with the best hearing ability, a significant advantage when it comes to capturing small rodents under snow or fallen leaves. Owls' half-moon-shaped ear holes are larger than those of other birds. These holes are asymmetrically arranged on each side of the owl's head with the result there is a slight offset between the two. This arrangement, combined with the owl's head movements when it detects motion on the forest floor, helps it locate moving objects in three-dimensional space.

Several owl species also have a parabolic facial disk that focuses sound waves towards their ears. Even the faintest sounds are picked up by these highly efficient nocturnal predators: the great-horned owl, for example, can pick up the sound of moving prey up to 330 m away. Thanks to the hearing part of their brain that contains more neurons than that of other species of comparable brain size, owls also process auditory information just as efficiently.

[14] Exceptionally, some bird species (notably pigeons, as well as some warblers) can hear sounds of 20 Hz or less. In fact, by tracking the movements of golden-winged warblers fitted with geolocator collars, researchers have discovered that these birds can anticipate and avoid bad weather by catching infrasound coming from even hundreds of kilometres away.

What's more, the plumage of Strigidae[15] (the most common owl family) has evolved over thousands of years to become perfectly silent: when a barred owl takes flight energetically flapping its wings, even animals with the best hearing can't hear it. This enables barred owls to surprise their prey by hearing, locating, and snatching them. So, can owls adjust their flight parameters as their prey moves across the ground? Do they make use of the Doppler effect?[16] We simply don't know.

Unlike for birds, hearing plays only a limited role in the lives of reptiles. For example, tortoises have poor hearing (in addition to being mostly mute, and snakes are deaf (as we generally understand the term), but they can still pick up low-frequency vibratory waves transmitted by the ground. Tortoises compensate for their poor hearing by drawing on powerful senses of sight and smell, while snakes use chemical and olfactory clues to detect everything in their habitat. The only reptiles that have an external ear are lizards, and of these, only geckos emit sounds to mark out their territory or court females.

The organ that provides amphibians (salamanders, frogs, and toads) with their sense of hearing has evolved from a sensory structure (the lateral line that can pick up sound waves in water) to an ear that can hear and respond to sound waves in air. This adaptation reflects the evolutionary transition of vertebrates, which have over time gradually extended their sphere of activity from water to land over time. Interestingly, while some amphibians still have a lateral line at the tadpole stage, this disappears by the time they become adults.

Male toads and frogs rely heavily on songs to attract females and reproduce. These animals hear via an eardrum on the fine, thin skin of their body surfaces, However, unlike the three ossicles in mammals,

[15] Name of the family of birds that includes most owls, with just over 260 species worldwide

[16] A physical phenomenon that explains the variation in frequency of a sound wave emitted by a moving body relative to a stationary hearer. For example, the sound of a police car siren is always higher-pitched as it moves closer to the hearer, but becomes lower-pitched as it moves farther away

this organ has only one ossicle to transmit airborne sounds to the inner ear.

The mammalian ear's three bony structures are an excellent example of the evolutionary transition of vertebrates. The ossicle (columella) of amphibians, reptiles, and birds is a structure derived from the same primeval fish jaw that became the stapes in mammals. As previously mentioned, the other two ossicles of the mammalian inner ear (the malleus and the incus), also derive from the jaw parts of the first vertebrates.

Since sound propagates differently through water compared with air and the ground, fish perceive sounds through a long sensory organ—the lateral line—on each side of their bodies. Water entering a duct via tiny pores in the epidermis of this line bends ciliated sensory cells. Comparable to the cilia in mammal and bird ears, this provides information about ocean currents and especially about low-frequency sound waves produced by nearby moving bodies, such as prey or predators. Is the sea a silent world? No, not really...

Invertebrates also have organs for emitting and capturing sound waves, which, as in the case of other animals, have evolved over time through natural selection. The sensory hairs on the bodies of many insects vibrate to certain frequencies. Tiny hairs on the antennae of male mosquitoes, for example, detect the whirring wings of flying females. In their case, any air movement generates a difference in pressure, however slight, which forces antenna hairs down to the roots, to the sensory structure (Johnston's organ) that detects the phenomenon. On the other hand, female common fruit flies use their Johnston's organ to recognize a male and, ultimately, be seduced by his sexual antics. The same mechanism operates in beehives whereby worker bees returning to the hive perform elaborate dances involving variations in the vibratory frequencies of their wings. These dances provide the other bees with valuable information on flying to the newly discovered sources of nectar. Small beetles of the Girinidae family also use a similar mechanism to move about on the surface of ponds and pools by comparing their own undulations on the water with those of other individuals in the vicinity. This is how they avoid collisions with other creatures, including other beetles, just like flying bats that navigate by echolocation.

Many insects have other hearing organs in unique places: crickets have tympanic membranes on their two front legs and cockroaches have organs sensitive to sound vibrations on each of their six legs. Notably, moths have tympanic membranes under their wings, behind which small air sacs, like soundboxes, transmit vibrations picked up by the outer membrane. In this way, many moth species have adapted to nights spent trying to avoid predators (especially bats) by detecting the ultrasonic waves emitted when they're being chased, and initiating erratic flight patterns to confuse their enemy. When luna moths, one of our most beautiful Lepidoptera species, are pursued in flight by a bat, they swirl the long tails hanging from their hind wings to alter the waves emitted by the stalking chiropteran. As a result, the bat attacks the moths' tails instead of their bodies, which gives the moths a better chance of escaping with their lives.

However, sound in invertebrates is not always produced by a specific organ. For example, green lacewings, translucid insects of the order Neuroptera, shudder with their whole body in an effort to find a mate amidst plant foliage. These vibrations propagate down plant stems and are thus easily picked up by other insects in the vicinity.

We can also take water boatmen[17] as another example. These pond and riverbank insects of the Corixidae family (of the same order Hemiptera as aphids, cicadas, and bugs), attract females by rubbing their front legs against their heads to produce seductive chirps.

[17] Water boatmen use their enlarged, spoon-shaped legs to move around their habitat of shallow riverbank water.

6

Battle Cries

As mentioned, many animal species use songs, calls, and other sounds to demarcate and defend their nest or territory during the breeding season or to attract a mate.

Tolerating the presence of other animals varies according to species and depends on factors such as the abundance of habitat resources, and the availability of mates. For most mammals and birds, the extent to which a nesting member of one species accepts other individuals of the same species diminishes the closer others get to the nesting member's nest.[18]

The black-backed woodpecker, which inhabits large coniferous stands, likes to frequent burnt forests, where it feeds on the insect larvae growing in charred tree trunks. Since such habitats are localized and ephemeral within the vast expanses of the boreal forest, it's not surprising that one pair of black-backed woodpeckers will fiercely defend their territory against not only another pair of the same species, but also against other woodpecker species, such as the American three-toed woodpecker, which frequents the same environments. To defend its territory, the black-backed woodpecker will patrol its stronghold,

[18] 'Home range' and 'territory' are concepts that are sometimes confused: home range is the area occupied by an animal to meet all its needs for food, water, and a sheltered space for reproduction; territory is a usually more restricted portion within the home range, i.e. the perimeter defended by a male or a couple against other members of the same species.

emitting strong, clipped *clicks*, and drumming vigorously on any tree trunk that can serve as a sounding board. Its drumming is faster and lasts longer than that of the striped-backed woodpecker, which differs from that of the American three-toed woodpecker by a sudden acceleration, like a ball bouncing on the ground. The defended territory in this case tends to correspond to approximately the area of the home range the black-backed woodpecker needs to feed itself and its offspring.

In the case of Northern gannets, on the other hand, there is a high level of tolerance towards other gannets in their home range's feeding areas, given that there are bountiful resources in the sea. Northern gannets are often seen in groups, stalking dense shoals of mackerel, capelin, and herring for food. However, this species does not hesitate to use aggressive noises and powerful blows of its beak to defend its nesting area and its offspring against any neighbour, young or old. However, the protected area is limited to the nest site and the area required for three individuals—the two parents and their offspring—to live together.

The northern short-tailed shrew is just as aggressive when it comes to a female defending her offspring or a male defending his mate. Confrontations, fights, threats, and shrill, warlike cries are thus commonplace during the breeding season. The belligerents emit rapid buzzes and chirps, while sometimes tapping their front legs on the ground. However, this shrew species is not concerned about defending its home range beyond the restricted perimeter of the nest. Like bats, it can emit echolocation calls in registers ranging from 30–55 kHz, which it uses to get its bearings when moving through underground tunnels.

Especially in late summer, the highly territorial American red squirrel stubbornly defends a roughly circular forest area of some 7,500 m^2, at the centre of which lies its main food reserve, often a pile of cones. After all, this war treasure has been amassed over many months and serves as the squirrel's reserve for the winter. Any intruder approaching the centre of the area is warned, first by an unpleasant rattle-like screech that can be heard over a distance of about 150 m. Then, if the message isn't understood, the squirrel lets out a series of high-pitched, screeching *tsoos!* followed by growls. Finally, if the intruder persists in approaching the central area, it is ruthlessly driven away.

The red-winged blackbird is another species with an over-developed territorial instinct. Reigning unchallenged over 300–1,000 m^2 of cattail marsh or wetland invaded by common reed,[19] the male of this species, which normally has a small harem of two or three females who lay eggs and brood within the same territory, fiercely defends a perimeter demarcated by its mating and nesting activities. With characteristic, hoarse *conk-la-ree!* cries emitted at regular intervals, he makes it clear that any male competitors of his own species, as well as potential predators like mink, racoon, or crow, are not welcome in his fiefdom.

Just as aggressive as the blackbird is the greater yellowlegs. Perched high on his watchtower, often the top of a stump or the shriveled top of a black spruce, the male of this species tirelessly watches his territory, standing guard over his very own patch of boreal bog. He proclaims his ownership rights with a resonating and tirelessly repeated refrain of whistled syllables: *Too-wee! Too-wee! Too-wee!* This vocal demonstration not only discourages any attempted intrusion, it also diverts the attention of potential predators and helps keep secret the nest site, a simple shallow depression in the moss, lined with twigs, sphagnum moss, leaves, and grass shreds, where the female broods in silence.

After hooded seals have spent the winter at sea, they gather for the mating season. Rutting males engage in endless disputes to gain access to a female. This is followed by grunts, roars, and other vocal or visual demonstrations directed at anyone attempting to replace them with their mate. These seals can inflate a cap of skin on their head, stretch-

[19] Or phragmites, an exotic strain of grassy plant that has become highly invasive in North America

ing from snout to forehead, into which they blow to make it grow to the size of a soccer ball. To discourage rivals, they also have a bright red stretchable nasal sac, which expands at will.

The barred owl is the most talkative of all the owl species in North American forests. Pairs, united for life, inhabit the same territory, season after season, year after year. The rare commodity in its habitat is not the quantity of food—the small mammals on which it feeds (moles, shrews, mice, and voles) are generally abundant in forest environments—but rather the availability of sufficiently large cavities in trees to accommodate a nest. It's hardly surprising, then, that a pair's territory, which averages one to two square kilometres during the breeding season, is vigorously defended by the male, and sometimes by the female as well, especially through use of the species' characteristic hooting often interpreted as: *Who cooks for you? Who cooks for you all?*

Grey treefrogs and ground frogs also vocalize loudly to set themselves apart from competitors and to attract females. In spring ponds, groups of green frog males inflate and deflate a vocal sac under their throats to emit dry clicks and powerful, guttural *toogs!* like a plucked banjo string. Each male takes advantage of brief moments of silence to express itself: the smallest emit in high registers, the largest in low ones.

Many orthopteran insects, such as crickets and grasshoppers follow a similar procedure when males try to mate with females. The male listens carefully to the position and vocalizations of other suitors, then slips their song into intervals whenever his rivals fall silent.

7

LOVE SONGS

In the evolutionary process of sexual selection,[20] male animals looking for mates are driven to set themselves apart from the many other males by showing off their prowess, including their ability to make powerful sounds.

In a well-known example, the male humpback whale's courtship ritual during its wintering seasons in the Southern Hemisphere is to demonstrate his vocal prowess as proof of his physical strength and the superior quality of his genes. This is because female humpback whales are instinctively looking to pass on the best genetic heritage to their offspring. The male humpback whale's song, lasting from a few minutes to over an hour, is divided into complex themes and captivating musical phrases that are superimposed and repeated as in a work of classical music. The frequency of this whale's sounds ranges from 200 Hz to 2,500 Hz, a vocal range equivalent to eight octaves in humans, and an intensity range of 155–189 dB. There are unexpected variations in the songs, depending on the isolation of the populations within the wintering areas. As seasons go by, certain songs produced by males from

[20] Sexual selection is another natural selection mechanism whereby the genes responsible for certain sexual traits (such as antlers in deer or feathers in peacocks), supposedly proof of exceptional health and outstanding physical vigour, are favoured and passed on to the next generation.

other areas seem to be more successful than others, somewhat like pop songs on the radio.

Of course, beauty is very much in the ear of the listener. So, what then of the hermit thrush's mating song? This fluid melody, repeated in an increasingly high register, is one of the most characteristic songs heard in the forests of northeastern North America. At just the right moment 40–45 minutes before sunrise, males that have arrived a few days or even weeks before the females in order to choose a territory on the breeding grounds begin their concert, tirelessly repeating their call until an interested partner shows up. Even after a pair has formed and the nest has been built, the male continues singing to defend the boundaries of his marital territory against other rivals in the vicinity. Hermit thrushes and Swainson's thrushes are among the few passerines that also sing at dusk.

However, when it comes to the amount of time spent in singing, thrushes can't hold a candle to the red-eyed vireo that sings at dawn, noon, and mid-afternoon. Several years ago, a single male of this species racked up an impressive total of 22,197 calls during a ten-hour period!

Male sharp-tailed grouse demonstrate the same kind of behaviour to attract females during their breeding season in late April or early May. In predetermined spots that have been used year after year for generations, the males dance, spread their wings, and show off their finery of orange-yellow wattles on their heads and featherless purple bags on either side of their necks. Within small areas just a few metres in diameter, they ruffle their feathers as they cluck, cackle, and intone low *coowoos!* at regular intervals. Interested females wander from male to male until one of them is seduced by a particularly attractive male. The match is quickly arranged and the pair fly off to make love away from prying eyes, while the other males continue to prance around until all the available females are taken up.

As with humpback whales, certain bird species show distinct patterns in their mating calls. In North America for some time now, the full version of the white-throated sparrow's famous call, *Old Sam, Peabody, Peabody, Peabody!*, has been heard less and less frequently with males curtailing the final note to come up with *Old Sam, Peabo, Peabo, Peabo!*—a version that is proving to be more popular with females than the full version!

So, how did this new version spread? This is thought to have happened on the wintering grounds where northern North America's populations of white-throated sparrows congregate together on the plains of Texas and Kansas. Buntings from the west, with the new version of the call, seem to have been the first to arrive at the wintering grounds, and from winter to winter, more and more buntings from central and eastern North America have preferred the new version.

The white-throated sparrow has two gene pools that give rise to different-looking sparrows: white-browed and buff-browed. Males of the white-browed variant are more aggressive and sing more than their buff-browed counterparts. They are also more fickle, mate more frequently with more than one female, and secrete higher levels of testosterone. In comparison, buff-browed males are better fathers and play more of a role in raising their young.

The mating songs of waterfowl (geese, ducks, etc.), shorebirds (plovers, sandpipers, etc.), birds of prey (buzzards, sparrowhawks, etc.), gallinaceous birds (grouse, etc.), and some passerines (flycatchers and other tyrant species), are somewhat innate, fixed, and similar, whereas those of most passerines, including buntings and mockingbirds, are more variable and reflect the acquired knowledge of the typical dialect the young learn by listening to their parents or their relatives around the nest.

Female red mockingbirds, as with other females of the mimid species, prefer males with the broadest vocal repertoire. At just the right moment, males try to impress passing females by unleashing their repertoire of songs and other noises (even imitating mammals, other bird species, and sounds like bells, etc.). With over 1,000 different musical stanzas recorded (some biologists put the figure at over 2,000), the male red mockingbird has one of the largest song repertoires of all birds. Each stanza is usually repeated twice, making it easier to identify the species when the bird is hiding in vegetation.

The grey treefrog, a small amphibian barely six centimetres long, is an arboreal species. To attract the attention of females that appear in May and June on the edges of forest ponds and spring pools, the males emit a rapid trill of about a second in length, which sounds like the ring of a telephone. These calls become more insistent at the height of the mating season, with a male uttering 20–22 trills per minute. When several of these frogs sing in chorus near the same spot, it's hard for a predator to identify the sole source of the singing. In this way, amphibian choral singing enables each chorist to blend into a kind of collective anonymity and protection that often, but not always, discourages predators. Grey treefrogs are loyal to their breeding sites and return to them year after year, without confrontation, especially since males generally cohabit peacefully.

Not so for the plainfin midshipman, a Pacific fish species whose males build nests on rocky ocean bottoms to attract females. Inter-male competition is intense, with some males singing to attract females to their nest. Only males with more developed brains sing by uttering curious guttural growls, while their mute counterparts are opportunistic and happy to wait nearby for a female attracted by the calls to lay her eggs. Once the eggs are laid, the cunning opportunists rush in to fertilize them by drenching them copiously with semen. Since the volume of sperm produced by the mute males, who have oversized testicles, exceeds that of the singing males, the former have a better chance of reproducing even though they had little to do with attracting the females.

Thanks to the work of a Japanese-Danish research team published in 2009,[2121] we now know that some male moths emit ultrasounds to

[21] R. Nakanao, T. Takanachi, T. Fuji, N. Skals, A. Surlykke, and Y. Ishikawa, "Moths are not silent, but whisper ultrasonic courtship songs," *Journal of Experimental Biology* 212 (24), (2009): 4072-4078.

attract females. These ultrasounds are produced in response to pheromones, the chemical sex signals emitted by females. These signals are very low-intensity, somewhat like whispers, to avoid giving away the sender's location to bats, the moths' archenemy.

In northeastern North America, the breeding season for singing insects extends from late July to mid-September. At dusk, crickets, grasshoppers, and cicadas start chirping a nocturnal chorus of continuous noise. The males of a relatively common bush-cricket species in southeastern Canada—sword-bearing coneheads (*Neoconocephalus ensiger*)—emit loud chirps to attract females. This jeopardizes their own lives because the calls also attract the attention of little northern long-eared myotises. Nevertheless, the bush-crickets can hear and counter their predator's echolocation calls by rhythmically emitting vocalizations, alternating between 30 milliseconds of stridulatory sound and 40 milliseconds of silence. When the bush-crickets detect ultrasonic pulses during their silent period, they immediately go quiet for several minutes. This strategy is effective as bats usually leave the area after the bush-crickets fall silent.

Of the fifty or so insect species that sing at this time of year, some, like the snowy tree cricket, are discreet and not well-known. This species, soft green in colour, motionless, and cleverly concealed in low vegetation, is considered the ghost of the night. Amazingly, its stridulation, calibrated to air temperature, is so regular that one can deduce the ambient temperature: simply count the number of pulses emitted every seven seconds and then add five to obtain the required Celsius value to the nearest degree. For example, if 14 pulses are emitted in 7 seconds, the ambient temperature is 19°C. The songs of males in the same area tend to synchronize with each other, amplifying the power of the whole and further regulating the rhythm.

✦

According to a hypothesis formulated by bioacoustician Bernie Krause[22] in the early 1990s, the characteristics of the acoustic signals emitted by animals in their habitat (frequency, intensity, duration, and timbre) are the result of natural selection during the breeding season. To make themselves heard in the cacophony of dawn, species that vocalize to find a mate use a distinctive register to minimize being drowned out by the calls of others.

According to Krause, amphibians, birds, mammals, and singing insects vocalize in specific frequency registers, in bandwidths chosen to contrast with others in the same way that a symphony orchestra's different instruments harmonize with the whole while remaining perfectly distinct. A good example is the clarinets in Stravinsky's *The Firebird*, or the double basses in Mozart's *Requiem in D minor K. 626*. In this way, frequencies and timeframes are less likely to overlap and mask each other.

The musical scores of animals species also take habitat geophony into account. For example, wolves and coyotes howl at dusk or at night, the times of day with fresher air, calmer winds, and less noise, so that their howls can be carried farther. Similarly, hermit thrushes and Swainson's thrushes often sing close to lakes and other calm waters at sunset, so that their monotonous chants can carry farther.

The American dipper is a small, dark grey, western passerine that superficially resembles a wren. Unlike the latter, which is more terrestrial, the dipper lives near mountain streams, where it feeds on aquatic invertebrates by diving headfirst into the current. Its calls and songs are astonishingly powerful in how they pierce the roar of waterfalls and rapids.

[22] American-born Bernie Krause started out his career as a studio musician, then became a sound engineer. During this time, he collaborated with many artists, including The Doors, and designed sound effects for films (*Apocalypse Now*) and TV series (*Mission Impossible and Bewitched*). Initially interested in experimental electroacoustics, he subsequently made hi-fi recordings of the sounds of nature and obtained a PhD in bioacoustics.

8

CALLS OF CONTACT, ALARM, AND DISTRESS

Animal species also vocalize for reasons other than defending a territory or finding a mate. These signals include calls among members of the same species moving in a group, calls from young to their parents for food, alarm calls to warn one or more band members about a threat within the habitat, or loud cries of distress in the event of an actual or suspected attack.

When on the move, individuals in a grey wolf pack use distinctive howls to maintain constant audible contact with each other.

Each animal in the pack has its own voice, which lets others locate it several kilometres away when conditions are favourable—for example, in open landscapes or when there is no wind. These calls are believed to be a means of coordinating the pack's movements, especially when hunting.

Male and female scarlet tanagers, a species that lives in mature deep forests, vocalize in their nesting habitat's dense undergrowth. Although visual contact between mates is difficult and females are rarely visible, they can locate their mates at all times from the sounds they make. The same goes for rose-breasted grosbeaks, a species that nests in young, regenerating forests. Since it is difficult for mates to see each other in such closed environments, the males and females tend to sing together.

For many bird species, once young chicks are freshly hatched from the egg, they remain in the nest for a few weeks to be fed by their parents. Both males and females do not skimp on the search for food, incessantly making several round trips an hour to supply their brood with insects, worms, and other protein-rich invertebrates. Every time a parent returns to the nest, it triggers cries and jostling in the chicks, who are in direct competition for parental attention and the best rations.

This parental feeding sometimes continues for several days after the young have flown from the nest but are still learning how to feed themselves. During this period, young long-eared and great-horned owls utter particularly shrill calls.

When young deer mice sense a potential danger, usually an approaching predator, they warn their mothers by issuing a variety of alarm calls, some in the ultrasonic range inaudible to humans and possibly also to predators. Young flying squirrels also have a high-pitched call that is beyond the range of human hearing.[23]

The eastern grey squirrel's alarm and distress calls warn both parents and fellow squirrels of threats. These calls are codified to indicate the level of perceived danger: a series of *hooks* indicates low level; nasal *quahas* or *bazzes*, medium level; and moans, high level.

When the black-tailed prairie dog, a colonial rodent that builds underground tunnel networks in the open spaces of western North America, spies a golden eagle, it goes into a state of high alert. If the eagle gets too close, one of the adults scanning the horizon from their mound of earth will let out a loud alarm to warn their burrow neighbours. The prairie dog has a repertoire of a dozen barks and calls with each tailored

[23] Researchers have hypothesized that the ultrasounds emitted by young rodents are not intentional but rather a physiological and mechanical reflex or side-effect caused by a sudden compression of the abdomen due to factors like extreme cold or behavioural stress.

to the predator's identity and threat level.

There are other settings in which animals vocalize, notably to warn away potentially attacking pursuers or enemies. When a confrontation seems imminent, several species resort to vocalizations to avoid a fight. For example, when long-tailed weasels are trapped, they open their mouths wide to emit an abrupt bark, suddenly advance towards their adversary, and then stand stock-still.

When black bears are surprised by another bear, they hiss, gnash their teeth, and stamp on the ground or on a tree trunk to warn off their rival. If the situation escalates, they growl and pretend to attack before retreating, while continuing to assess their opponent's strength.

Although snakes of the *Crotalus* genus (Viperidae family) are deaf, they possess a rattle at the tip of their tail. This is a unique structure the snakes shake to produce a loud, rattling sound that indicates their presence in a conflict situation. This organ is composed of a conglomeration of scales from previous molts that form layers of horny, interlocking, mobile rings. Sound is produced when the layers rub together. Other snakes within the same family have specialized scales along their flanks: when these snakes curl up, the scales rub together and make a powerful screeching sound that is capable of frightening off any potential enemy.

Fourth Movement

Plant Soundscapes

Trees spoke before humans did.

Joséphine Bacon
Bâtons à message

Elms are not as mute as you might think.

Frère Marie-Victorin
Croquis laurentiens

9

Trees are Sentient Beings

In the same way that animals use their senses to obtain information about their environment, plants in general and trees in particular use certain mechanisms to get information about their surroundings.

As in animal species, natural selection has endowed plant species with structures and functions that help them survive and reproduce in the best way possible by coping with changes in their habitats.

Plants can detect changes in humidity, temperature, luminosity, and chemical composition in soil and air. This enables them to compensate for unwelcome changes and take advantage of favourable ones.

A striking example of this is the ability of trees in temperate zones to monitor for signs of the changing seasons. By detecting certain factors, notably daylight length (photoperiod) and other seasonal phenomena like severe late-summer drought, hardwoods such as striped maple, sugar maple, and swamp birch know when to go dormant before winter and when to emerge from dormancy in the spring.[24]

Unlike animals, plants cannot move if their immediate environment is no longer suitable: they are literally rooted to the spot where their seed germinated. Plants must therefore be able to sense and process

[24] Dormancy enables plants to avoid long, critical periods when water in liquid form is unavailable due to sub-zero temperatures.

environmental stimuli, and anticipate future constraints in terms of fewer resources (water, minerals, or light). Plants can use various hormones[25] to modify their responses to environmental stimuli. Two of these hormones, ethylene and auxin, trigger leaf shedding in the fall, while another, gibberellin, triggers the post-dormancy emergence of plants in spring.

While animals possess a central nervous system to process information from the outside world, plants have no equivalent because their sensory organs—roots, trunk, branches, and leaves—are decentralized.

Surprisingly, plants have truly remarkable sensory capacities not at all inferior to those of animals. In their own way, plants see, touch, and smell.

See

To achieve photosynthesis and produce carbohydrates for growth, plants need three ingredients: water, light, and carbon dioxide.

Light also provides plants with information about the day-after-day and season-after-season passage of time. Plants sense not only the presence of light, but also its intensity, position, and wavelength.[26] Plants capture solar radiation at wavelengths of 400–700 nanometres (nm) in the visible light spectrum.

Blue light photoreceptors, stimulated by solar radiation at wavelengths of 450–500 nm, are the basis of phototropism, the natural tendency of plants to turn and grow towards light. These photoreceptors also trigger the opening of stomata, the microscopic pores on leaf surfaces through which plants transpire excess water from their tissue.

Other photoreceptors in the red light spectrum (600–800 nm), known as phytochromes, provide plants with information about light

[25] Phytohormones

[26] The human eye perceives different wavelengths as colours.

quality so that they can adapt to luminosity variations. The red light that reaches the undergrowth on the forest floor under the tree canopy is weak because much of the light red radiation (wavelengths of 600–700 nm) is intercepted by canopy leaves. Far-red light (wavelengths of 700–800 nm), on the other hand, reaches the ground more easily. The ratio between the two types of red indicates to a below-canopy tree that it needs to devote more energy for its terminal stem to grow taller. Conversely, the same tree in full sunlight receives more red light in the 600–700 nm range, a sign that it needs to branch out horizontally and slow its upward growth.

Phytochromes also play a role in triggering seed germination with germination being stimulated by more red light than far-red.

Light quality also controls plants' natural rhythms of activity and rest, in the same way that it affects animals' sleep-wake cycles. These circadian rhythms, which occur in all living organisms, follow a periodicity of around twenty-four hours. Since plants are unable to produce energy without light, they slow down their tissue-level cellular metabolism at night. The changing proportion of different wavelengths during the day therefore has a direct effect on plants' biological clocks: when the proportion of red light in the spectrum is high at sunrise, the photosynthesis factory starts up; when the reds become more intense once again at dusk, the factory shuts down till the following morning.

These two types of plant photoreceptors—those that respond to blue light and those that respond to red light—act in the same way that the eyes of animals capture and route stimuli in the same wavelength spectrum so that the animals can adjust their behaviour.

Animals and plants utilize the same photons from the sun but have chosen different evolutionary routes to make use of this universal energy.

Touch

Plants don't only react to light; they also modify their architecture and behaviour in response to other stimuli, including those that indicate physical constraints in their immediate environment.

Who hasn't seen a tree wrap its trunk around a fence or an old clothesline pulley? As soon as the tree's living tissue encounters an inert, foreign object, cell growth is stimulated to form a kind of mantle around the object. The same phenomenon occurs when trees' above-ground roots encircle or outflank large rocks. The evolutionary origin of this process is clearly to improve trees' mechanical strength against threats from gravity or climatic elements (wind, ice, etc.). For the same reason, trees will sometimes anchor themselves to the branch of another tree, whether of the same species or not, while two underground roots of the same or another tree will also become entwined with each other.

The trajectory of a tree's roots beneath the ground is determined by how the root cap, the root's "seeking head," responds to obstacles like stones or other roots as the tree develops its root network.

The above-ground parts of plants are also sensitive to touch: if the stem of a young plant is rubbed a few times a day, it develops less well than another plant that has not been touched.

Some plants, such as mimosa, respond to physical stimuli by rapidly moving their foliage. A second or two after their leaflets are touched, these fold back to expose small spines on the plant stem that discourage herbivores from feeding on them.

Over evolutionary time, some species have also developed the ability to use this sense of touch in other ways. For example, prehensile tendrils on the stems of vines and other climbing plants enable these plants to cling to supports and grow towards light at other species' expense. The small structures emerging from the stems grow in a straight line until they meet an object. When this happens, the structures' cells start to grow at different speeds, depending on which side they're on, in a way that bends and transforms them into efficient hooks. This phenomenon is called thigmotropism.

Insectivorous plants like sundews (a small species that grows in wetlands) have sensory hairs on their specialized leaves that trigger rapid closure of the leaves whenever insects land on them.

Smell

In response to the constant threat of herbivores, whether large like deer or small like defoliating insects, plants have developed mechanisms to defend themselves. Such mechanisms include both physical features (hair and spines) and organic chemical compounds that are toxic or repellent.

Plants recognize their natural enemies' chemical signatures and can differentiate a host of compounds like oligosaccharides, peptides, and enzymes. For example, plants can detect the regurgitations or oral secretions of caterpillars that graze on them. They can also sense when tiny female moths lay eggs on their leaves because the substances deposited with the eggs modify leaf tissue chemistry. The latter situation triggers a local defensive reaction that makes plant leaves less conducive to the birth of more grazing caterpillars.

When attacked by herbivores, plants can respond by simultaneously emitting from as few as 20 to as many as 200 types of volatile organic compounds like methyl jasmonate or methyl salicylate. These self-defence molecules, which circulate faster in air than in plants' vascular tissue, enable the parts located near the attack location to rapidly obtain information and react by producing more repellent or indigestible substances.

These alarm signals can be detected within radii of 60 cm–10 m around the emission site, and even farther if the wind is blowing in a particular direction. The molecules emitted into the atmosphere by the attacked plant act as chemical warnings for neighbouring plants, which then produce more repellent molecules, such as ethylene and jasmonic acid, to make them less desirable to hungry herbivores.

Some even more ingenious plant species release volatile compounds that attract herbivore enemies. For example, corn leaves fend off invasions of noctuid moth caterpillars by releasing highly volatile substances that attract parasitoid wasps that lay their eggs directly in the caterpillars' bodies.

Birds are just as sensitive as many insects to the chemical composition of substances released by plants. For example, small forest sparrows in Europe are attracted by the gaseous compounds released by downy birches when moth caterpillars attack their foliage. The compounds act like a cruise ship bell announcing that dinner is served!

✦

If plants can see, touch, and smell, can they also hear by receiving and processing sound waves? Recent research has quite a lot to say on this topic.

10

How Plants Hear

We still know very little about bioacoustic phenomena in the plant world because this is a new research field. Nevertheless, researchers are beginning to discover some remarkable facts.

For a start, the ability to perceive sound saves plants considerable amounts of energy.

Sound waves and other vibrations in the soil provide plants with valuable information for detecting cavities, resurgent water, or trickles of water beneath the surface, and then directing their root systems to find the moisture needed for plant maintenance and growth.

In 2017, four researchers led by Monica Gagliano, a professor evolutionary ecology at the University of Sydney in Australia, showed that the roots of peas, an annual species in the Fabaceae family (legumes), orient themselves and find water by tracking vibrations produced by water movement in the soil. Young corn shoots showed the same behaviour in response to a sound source at 220 Hz.

The intensity of signals emitted and perceived by forest plants is highest in the rhizosphere, a zone located a few metres below the soil surface. The intertwined root systems of trees, shrubs, and herbaceous plants extending in all directions permeate this narrow zone full of fungi and other microorganisms, and benefit from their association with the fungi's hyphal system to create highly efficient interconnected underground networks.[27]

Like the traffic characteristic of densely-populated, vehicle-clogged cities, the rhizosphere is the site of uninterrupted flows of chemical, electrical, and mechanical messages.

✦

As described in previous chapters, natural selection in many vertebrates (mammals, birds, and amphibians) has favoured the development of external or internal structures for capturing sound waves. These structures are essentially capable of transforming acoustic or vibratory energy into mechanical energy. Although natural selection has operated differently for invertebrates, the hearing outcomes have remained the same. Mosquitoes and fruit flies hear with their antennae; cockroaches with each of their six legs; crickets with eardrums located on their front legs.

Other animal species, such as grass snakes, are completely deaf in the normal sense of the word, but can still use their jawbones to pick up vibrations transmitted through the ground.

The physiological mechanisms plants use to detect and respond to

[27] The symbiotic association between fungi and plant roots is called mycorrhiza. This association is believed to have helped the first plants colonize land more than 400 million years ago. The symbiosis benefits both parties: in exchange for the sugars produced by the plant through photosynthesis, the fungi provide a complementary supply of water and other essential nutrients (mainly phosphorus and nitrogen). One cubic centimetre of forest floor can harbour up to 100 m of fine, branched hyphae.

sound waves are poorly understood, but we now know that plants pick up and interpret vibratory signals circulating beneath the ground. Still, questions abound. How far can roots detect vibratory waves? What is the minimum perceptible intensity threshold? Are some frequencies more likely than others to stimulate roots to look for a water source?

✦

Plants live between two worlds, one above ground (air) and the other below (soil), and simultaneously draw essential elements from both: carbon dioxide and solar radiation from the air, and water and other nutrients from the soil. Plants also gather information from both above and below ground.

Plants also pick up sound waves from their own surfaces as well as via the soil. In a striking example, specimens of thale cress, a member of the brassica family, were exposed in a lab to the sounds of a moth caterpillar munching on a leaf. Plants exposed to this sound released harmful molecules in their foliage to repel the supposed assailant.

When exposed to other types of natural sound (wind-rustled leaves or the songs of other insects), the plants in the same experiment did not react with a perceptible chemical response.

We still have a lot more to learn from plant acoustics. In another remarkable experiment, a team of eighteen researchers at Tel Aviv University, Israel, discovered that large-flowered evening primroses change the composition of their nectar in under three minutes when they sense the buzzing of nearby butterflies or worker bees. By raising the sugar content of their nectar (by up to 20%) to make it more attractive, the plants hosted longer visits from insects on their flowers and thus more pollen sticking to the insects' legs. The plants were thus more likely to pollinate other members of their species and pass on their genetic heritage to the next generation.

Why don't evening primroses spontaneously produce sweeter nectar all the time? Because this requires a lot of energy, and it would be a waste of energy to do so if there were no foraging insects around.

The production of sugar in the nectar of evening primroses is linked to the sound frequencies of pollinating insects' buzzes: in the Israeli experiment, only evening primrose plants exposed to low frequencies of 0.2–0.5 kHz produced more sugar for their flowers.

Attracting pollinators efficiently is a key issue for flowering plants, given that the vast majority (87.5%) depend on animal pollination for reproduction.

The shape of the flowers must be not only inviting to insects, but also play a role in the plant's ability to capture and absorb the airborne vibrations generated by the wings of butterflies, bumblebees, and honeybees. For example, the concave, parabolic antenna-type shape of many flowers helps to focus sound waves. After the researchers removed one or more petals from certain evening primrose flowers, they found that the plants were less receptive to the buzzing frequencies of foraging insects. On the other hand, whenever the foraging insects' wings vibrated petals, a rapid behavioural reaction was triggered that concentrated the sugars in the plant's nectar more effectively.

✦

In the 1970s, a few pseudo-scientific stories persuaded some people that plants were sensitive to "energy fields" or particular "vibrations." This even led some individuals to conclude that plants prefer classical music to pop or rock because they had observed higher growth rates when plants were exposed to music by Chopin, Mozart or Sibelius. To date, however, no serious, reproducible experiment published in a scientific journal has confirmed such preferences.

While plants are indeed sensitive to sound, they don't differentiate noise from music. Natural selection is the reason why plants perceive sound waves. In other words, since there is an evolutionary advantage for plants to detect threats or a primary resource like water in the soil, such auditory capacities in plants are definitely real.

Could the tactile capacity of plants, as explained in the previous chapter, explain their sensitivity to sound waves? Possibly, because sound is a vibratory energy that travels through different media—soil, air, water, or plant tissue—and the lowest sound frequencies can cause physical bodies to vibrate.

Sound, like light, is a phenomenon whose mechanical components plants can capture and interpret in terms of positive or negative signals concerning their environment.

Acoustic vibratory energy is present in all natural environments and all types of habitat—for example, beside ponds and lakes, in woods and deep forests, on bare peaks, and in soggy valleys.

As animals with developed hearing, humans naturally assume that auditory communication requires specialized hearing equipment to capture vibrations. However, plants don't have such equipment, nor do rudimentary organisms like bacteria. On the other hand, as Japanese researchers noted in 1998, this deficiency doesn't prevent bacteria from sensing, analyzing, reacting to, and even producing sounds. The Japanese researchers demonstrated that the *Bacillus subtilis* bacterium emits sounds in a frequency range of 8–43 kHz, and that a close relative, *Bacillus carboniphilus*, picks them up. In their opinion, this transmission-reception of vibratory waves between bacteria at a microscopic level plays a role in regulating bacteria growth.

✦

Thanks to their ability to process multiple stimuli from the external environment, plants, like animals, are endowed with proprioception, the ability to perceive themselves in space and differentiate between what is integral to them (their various parts) and what is not.

Proprioception in animals involves the capturing of countless messages by muscle and ligament receptors, as well as by gravity-sensitive ears in the case of vertebrates, and then their transmission to a central nervous system.

Plants also sense gravity, with gravitational stimuli capable of changing plant architecture and behaviour. If you lay a young shoot on its side, over the next few hours its stem will slowly curve upwards, and its roots downwards. This is because plants can differentiate up from down thanks to statoliths, dense starch compounds in plant cells that migrate under gravity to the plant's interior and, by a simple mechanical law, cause the entire plant to obey gravity. This is why tree growth on mountainsides slowly straightens out vertically to capture more of the sun's rays.

In conjunction with the stimuli of perceiving red light and far-red light (as discussed above), a plant's ability to sense gravity also contributes to apical shoot elongation (i.e., upward development of the main stem).

The discovery that plants are capable of proprioception is disquieting: not only does it reveal that plants possess previously unsuspected capacities for interaction with their immediate environment, it also opens up profound questions about the existence of a kind of self-awareness in plants.

In its simplest definition, isn't consciousness basically a living being's recognition of the existence of an external world in relation to its own inner world?

Several plant biologists recently concluded that plants do possess a certain degree of self-awareness and can learn from past experience. These are surprising findings that deserve further explanation.

✦

Professor Monica Gagliano carried out an interesting experiment on the behaviour of peas, the plant mentioned at the beginning of this chapter.

Using the classical conditioning method that Russian physiologist Ivan Petrovich Pavlov applied to dogs in the late nineteenth century,[28] she devised an experiment in which pea plants were exposed to light wavelengths that plants particularly appreciate (somewhat like Pavlov's food variable for dogs). This produced an expected behavioural response: phototropism caused the peas to turn towards the light. Exposure to these wavelengths was also accompanied by the whirring of a small fan (conditioned stimulus). After several repetitions of the procedure, and with the light switched off, the pea plants continued to turn in the same direction while the fan was running (conditioned response). In other words, the peas seem to have anticipated that light energy was going to arrive shortly after the fan was switched on! Professor Gagliano thus demonstrated that plants can be conditioned to adopt certain behaviours.

✦

The hypothesis of extended cognition (HEC) is another conjecture drawn from the animal kingdom that could provide some clues about plants' remarkable sensory faculties.

This hypothesis, formulated in 1988 by American researchers Andy Clark, University of Washington, and David J. Chalmers, University of Arizona, postulates that an animal's cognitive processes (e.g., how they interpret stimuli or learn about their environment) sometimes extends beyond the physical boundary of their own body. To understand this

[28] In his study of the mechanism whereby dogs salivate in response to food in front of them, Pavlov observed that the dogs began to salivate as soon as they heard the footsteps of his assistant bringing them food. Pavlov then introduced the sound of a metronome just before feeding the dogs. After several repetitions of this procedure, he then used just the metronome. As expected, the sound of the metronome alone (conditioned stimulus) triggered saliva production (conditioned response) in the dogs. Pavlov repeated the experiment with a bell and found that any object or event the dogs learned to associate with food triggered the same behavioural response of salivation.

hypothesis, one need only think of how humans use a variety of technological tools, such as mobile phones and personal computers to give themselves immediate daily access to a wealth of knowledge via the Internet. A pencil and a sheet of paper are also cognitive tools outside our bodies that help us perform calculations, prepare to-do lists, and so on.

A spider's central nervous system also extends beyond the physical barrier of its body. Although spiders only have small brains, they transfer part of their cognitive processes to their webs, which serve as sensory filters that distinguish between vibrations caused by wind and those caused by prey caught in the threads of their webs. This differentiation enables spiders to react only when it's worthwhile to do so.

How is it possible for plants to have sensory abilities, just like animals, and simultaneously perceive several parameters (e.g., temperature, humidity, gravity, and volatile organic compounds emitted by nearby plants), without possessing a central nervous system or localized organs? According to researchers André Parise, Monica Gagliano, and Gustavo Souza,[29] plants use two processes:

(1) they release organic compounds that modify the rhizosphere below the ground surface and improve their own perception/detection threshold: for example, plants release certain fluids through their roots that help them sense underground obstacles like large stones;

(2) plants manipulate subterranean microbial communities by releasing organic compounds through their roots that tend to favour one type of bacterial community over others. The beneficiary community then marks its host soil with more plant-specific stimuli in the same way that webs continually inform spiders about what's happening in the neighbourhood.

[29] André Geremia Parise, Monica Gagliano, and Gustavo Maia Souza, "Extended cognition in plants: Is it possible?" Plant Signaling & Behavior 15 (2) (2020): e1710661. doi.10.1080/15592324.2019.1710661.

Let's be frank, yet humble: the plant world is so different from our own that our tools and methods for unravelling its mysteries are not yet very effective.

✦

Energy in the form of vibratory waves is present everywhere—from the microscopic world of atoms and molecules to the large-scale world of earthquakes, tsunamis, volcanic eruptions, and lightning storms.

The sensory ability of living organisms to perceive and use sound waves to their advantage is widely distributed throughout the animal world.

Let's be confident that research over the next few years will open up new perspectives on how plants, which were on Earth long before us, interpret the world around them.

FIFTH MOVEMENT

The Songs of the Trees

But now all is changed,
and not even the return of the birds may be taken
for granted.

Rachel Carson
Silent Spring

11

THE NEW GLOBAL SOUNDSCAPE

Although the semantic similarity between *soundscape* and *landscape* is a currently convenient way to illustrate certain concepts (as discussed in this book's introduction), it masks an important difference when it comes to reciprocal sensory impact.

While we can easily ignore unpleasant sights by simply closing our eyes, we can't do the same with annoying sounds because ears have no lids.

Unless you're deaf or can physically move away from intolerably loud sounds or oppressive background noise, you forced to put up with it.[30]

Unlike our senses of sight or touch, we have no control over our hearing (or our sense of smell, for that matter). After we've closed our eyes at bedtime, hearing is the last of our senses to shut down. And it's the first to wake up in the morning.

This means that our nervous system is permanently immersed in a sensory environment that is endlessly generating sound stimuli, many of them unwanted and artificial, especially in urban environments. *Homo sapiens* was not born with a jackhammer in his hands!

[30] A Nonetheless, the human ear can temporarily lower the noise level received by the inner ear by an involuntary contraction of two small muscles in the middle ear, a mechanism known as the stapedial reflex. This reflex is triggered by a loud sound of at least 80 dB. However, this muscular tension cannot be maintained for long, and after only a few seconds of exposure, it drops by around half.

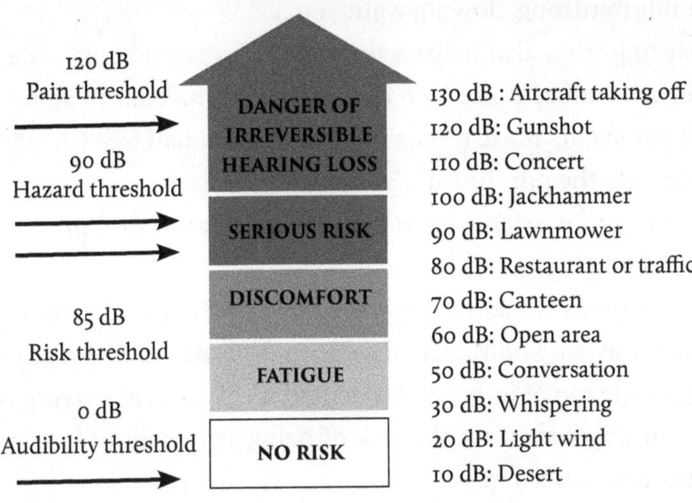

Table 4 Sound intensity is non-linear; it doubles every 3 dB. This means that a 15 dB sound is twice as loud as a 12 dB sound. The upper tolerance limit for the human ear is around 115–120 dB, equivalent to the sound of a gunshot. Prolonged exposure above 120 dB can damage human health (causing afflictions like tinnitus or temporary or permanent deafness).

The perpetual noise pollution we are subjected to all our lives—from sunrise to sunset from the cradle to the grave—affects our well-being, including our mental health. In environmental chemistry, insidious molecules toxic to human health are often described as nano-aggressors. By extension, can we refer to the hundreds of noises we encounter every day, especially in cities, as sono-aggressors?

These annoyances can be minimized by resorting to well-known strategies to counteract the extensive impact of man-made noise.

These strategies include erecting vegetation noise barriers and increasing the surface area of parks and green spaces, which encourages natural sounds (birdsong, flowing water, etc.).

Planting trees also helps reduce urban noise: a 30 m-wide barrier of trees can reduce noise by 8 dB, which is significant in so far as a 10 dB reduction cuts noise intensity by more than half (see Chapter 4 and elsewhere on the non-linear dB scale).

Paying more attention to soundscapes can only improve public health in cities.

The nervous tension experienced by city-dwellers trying to cross a busy, noisy avenue in Montreal or Toronto is not unlike that of a bird singing at the top of its voice in the hubbub of suburbia, trying to make itself heard by a mate, at the risk of being snatched up by a predator who has also noticed it.

In a humanized environment, the presence of natural sounds (biophony and geophony) can mask some of the man-made noise and improve the soundscape experience. Moreover, exposure to natural sounds, like waterfalls or river rapids, has been shown to improve stress tolerance and cognitive performance.

✦

Unlike visual landscapes, which are relatively stable over the course of the day (although they change with the seasons and, even more subtly from hour to hour as the sun travels across the sky), soundscapes constantly change from minute to minute, like a bird concert at dawn, which is never the same from one morning to the next.

Once again without our knowledge, our efforts to scan and make sense of this uninterrupted audio playback—this continuous series of soundscapes—in terms of threats or opportunities also causes sensory fatigue.

Who hasn't returned home exhausted in the evening after a long dinner in a particularly noisy restaurant?

✦

Out of curiosity to learn more about the soundscapes in my life, I acquired a good-quality sound level meter to measure the decibel-intensity of the sounds to which I am exposed in different places and circumstances at various times of the day. As part of my little experiment, I also noted the nature of the sounds in which I am immersed, and classified them into three broad categories: anthropophonic (traffic noise, planes in the sky, etc.), biophonic (birds, crickets, amphibians, etc.) and geophonic (wind, water, etc.).

I tested three types of setting: (1) my immediate environment—the house and its surroundings where I live, eat, sleep, and rest; (2) my local environment—the neighbourhood where I shop, walk, and ride my bike; (3) my work environment—the forests, marshes, lakes, and rivers I visit as a biologist and the in-person or online meetings I attend with my colleagues.

Soundscapes

Setting	Average dB	Anthropophonic	Biophonic (%)	Geophonic (%)	Remarks
Home					
Bedroom	34.4	85	15	0	Biophony: crickets (window open)
Dining room	37.4	95	5	0	Anthropophony: refrigerator compressor Biophony: blue jay
Garden	46	50	50	0	Anthropophony: train, plane, children's voices, construction site noise Biophony: birds
Neighbourhood					
Urban boulevard 1	64.7	100	0	0	Anthropophony: road traffic on a Sunday afternoon
Urban boulevard 2	69.1	100	0	0	The same site on a Wednesday afternoon
Highway	78.4	100	0	0	With a peak of 98 dB
Supermarket	64.2	100	0	0	

Work					
Office	**47.8**	**100**	**0**	**0**	Anthropophony: traffic (window open)
Forest (wild landscape)	**38.3**	**5**	**5**	**90**	Geophony: wind in trees Biophony: birdsong Anthropophony: jumbo jet at high altitude
Forest (agro-forestry landscape)	**41.2**	**25**	**5**	**70**	Geophony: wind in a jack pine Anthropophony: traffic in medium-distance Biophony: woodpecker
Forest (urban landscape)	**55.7**	**90**	**10**	**0**	Anthropophony: traffic Biophony: various birds
River	**57.1**	**0**	**0**	**100**	Geophony: wind and flowing water
River (rapids)	**69.5**	**0**	**0**	**100**	Geophony: wind and flowing water

Table 5 Noise levels in each setting were measured in the same month (September), under similar conditions (winds of less than 20 km/h, force 0–3 on the Beaufort scale, and no precipitation). The average dB figure represents the average of 10 sound level meter readings taken at 60-second intervals over a 10-minute period.

First set of findings: As someone living in the second suburban ring of a large city—Montreal—and at a considerable distance from a freeway, I'm one of those privileged few who aren't regularly exposed to high noise levels. For example, in the backyard of my house on a bright September afternoon, the average sound level over a 10-minute period was 46 dB, which is considered low. For the same length of time on the same afternoon, the sound level in the middle of the dining room inside the house averaged 37.4 dB, slightly higher than the 34.4 dB in the bedroom in the middle of the night.

Second set of findings: My neighbourhood isn't as noisy as I thought. Although the noise level of the highway, 1.8 km northwest of my home as the crow flies, is 78.4 dB, and that of the nearest urban boulevard, 1.4 km away, is 69.1 dB at rush hour and 64.7 dB at the quietest time of day, this is only slightly higher than the average value of 64.2 dB inside the supermarket where I shop. Since these are places I usually just visit, my exposure to these high noise levels is limited. And these values do not represent a threat to my hearing.

Third set of findings: I'm also very lucky to work in a quiet natural environment, largely free of man-made noise, except, inevitably, in the office. If it weren't for my profession, I'd have very little exposure to natural sounds. Furthermore, the sound intensity to which I'm exposed at work is low, whereas when I'm out in the field—in forests and occasionally near watercourses, my sound exposure ranges from 38.3 dB (backcountry forest) to 69.5 dB (rivers with rapids). Nevertheless, the table summarizing my limited experience shows that, even in deep forests, far from human infrastructure, it's hard to escape the sounds generated by *Homo sapiens*. All it takes is a jumbo jet passing overhead for the magical impression of being in a totally natural world to disappear. The only place where anthropophony completely disappears is on the banks of a fast-flowing river. When you're near rapids at over 69 dB, all you can hear is the roar of the water. Curiously, that soundscape is as loud as an urban boulevard at rush hour, but infinitely more soothing.

✦

During its prenatal life, between the twenty-fourth and twenty-sixth week, the human embryo begins to hear sounds such as circulatory and digestive noises and its mother's breathing and heartbeat. Hearing is the fetus's most acute sense.

The external sounds the fetus hears, such as voices, music, and various everyday noises, are medium- to low-frequency, since they are filtered down from high frequency by the mother's body fluid and abdominal walls as well as by the surrounding amniotic fluid, made up of 96% water. The average sound level of a fetus's soundscape is approximately 30 dB, which is equivalent to that of people whispering in a room.

The human auditory system is fully formed around the thirtieth-week of pregnancy. The cochlea, which houses the sensory hair cells (see Chapter 5), is fully functional by then. The number of these cells increases considerably after birth and during childhood, as does hearing acuity. A child begins to perceive high-pitched sounds, as an adult does, around the age of five, and the lowest-pitched sounds around the age of ten.

✦

From the moment we begin to perceive sound in the womb, noise never leaves us. We hear it throughout our lives, right up to our last breath. Complete silence is impossible unless we go for a walk in space!

Over the years, we're exposed to a bewildering variety of noises, in terms of timbre, frequency, duration and intensity. In addition, we gradually lose hair cells as we age and can also lose them because of accidental or intentional exposure to excessive noise for too long.

While man-made sounds are among the most powerful around us, some natural sounds can also be extremely loud. The loudest of these by far are volcanic eruptions. The sound generated by the eruption of the Indonesian volcano Krakatoa in August 1883 was the loudest sound ever recorded in human history. The eruption was audible as far away as Australia and New Zealand, and even as far as Indian Ocean islands 4,800 km away. Measured by the rise in barometric pressure, the sound intensity of the eruption would have reached 172 dB[31] at a distance of 160 km, i.e., a sound wave 16,000 times louder than the 130 dB[32] of an airliner at take-off.

The noise in the volcano's vicinity undoubtedly deafened many of the disaster's survivors.

For them, silence became a reality.

✦

The eye is at the top of the hierarchy of the five senses, at least in Western culture. As the organ of reason and objective knowledge, the eye strives to grasp what is elusive in the world. The ear, on the other hand, is undoubtedly the primary organ of apprehension and fear.

Always on the alert to detect potential threats and ward off danger, the ear feeds our mammalian brain, which is constantly trying to decode undefined noises and fragmentary information of unknown origin. The ear tries to make sense of these sounds, which constantly vary in duration and intensity, and which it situates more or less well in space.

✦

If we add up all the natural (geophonic or biophonic) or artificial (anthropophonic) decibels emitted each year on the planet, what is the most common and most constant sound humans hear?

If we only had one background track to give to extraterrestrial tourists to prepare them for their stay on Earth, what audio file would we send them?

The wind in the conifers of the great boreal forests that stretch

[31] Other sources indicate 180 dB.

[32] As mentioned in Chapter 4 measurements on the non-linear sound intensity scale double every three dB. This means, for example, that 15 dB is twice as loud as 12 dB.

across the northern hemisphere?[33]

No.

The sound of rain? Water ricocheting off the pebbles of the hundreds of millions of rivulets that converge into on all the world's seas? The crashing of waves on thousands of kilometres of ocean shoreline? The roar of the 750,000 to 1.8 million thunderstorms that erupt around the world every year?

No.

Most likely, it would be the sound of traffic. Cars and trucks passing by, horns honking, alarms going off at random, engines decelerating and then revving up again at green lights, the sound of compressed air brakes, and so on. These are the world's new soundtracks—the new global soundscape.

A growing body of research points to significant health risks for people exposed to traffic-related noise pollution. In the European Union, it is estimated that 22 million people suffer from various disorders linked to chronic exposure to such noise, including 6.5 million whose sleep is deeply disturbed. The noise intensity of urban traffic is responsible for 12,000 premature deaths every year.

According to the World Health Organization, chronic exposure to excessive noise levels increases the likelihood of mortality associated with high blood pressure, a major risk factor for certain diseases. In Europe, a study of over 57,000 people by a Danish Cancer Institute research team showed that exposure to noise 10 dB higher than average traffic noise resulted in a 27% increase in the stroke risk for people 64.5 years of age and older. The effect of traffic-related noise pollution can also be measured in real time. For instance, University of Madrid researcher Alberto Recio and his colleagues calculated the impact of night-time noise on the mortality risk for people in the same age group

[33] The coniferous forest areas of Russia, Canada and the Scandinavian countries form the largest continuous forest in the world.

(65 and over): each 1 dB increase in traffic noise at time of exposure raised the probability of a fatal heart attack by 3.5%, and of a fatal stroke by 2.4%.

A national health analysis in South Korea estimated that, for every dB increase in urban noise exposure, the number of cardiovascular or cerebrovascular accidents rose by 0.17–0.66%. This may not sound like much, but in a population of some 50 million, that represents an increase of 85,000–330,000 cases.

The individuals most at risk of suffering the after-effects of chronic exposure to high noise levels are seniors and young children.

Socio-economic status is obviously a factor in noise exposure: people with the lowest incomes are generally forced to live in neighbourhoods with anxiety-inducing soundscapes due to living near industrial zones, freeways, or other heavy traffic thoroughfares.

Conversely, a recent study by researchers from Canadian, American and New Zealand universities reported a 28% drop in stress, a reduction in various aches and pains, and an overall improvement in cognitive performance in people exposed to natural soundscapes in US national parks.

✦

In 1977, when NASA launched its two Voyager probes into interstellar space, each vessel contained a golden CD entitled *The Sounds of Earth.* Each CD featured soundtracks intended to represent the very essence of life on our planet for extraterrestrials. At the time, the most appropriate sounds deemed to evoke the fundamental elements of terrestrial human culture were sounds such as a mother and child kissing, a train moving down a track, and the sounds of humanity's first tools. Wouldn't it have been more appropriate to have reproduced the noise of constant traffic on a highway somewhere?

12

The Background Noise of Species Extinction

Ten erstwhile animal species in Quebec are now officially extinct: elk, walrus, cougar, brown bear[34], passenger pigeon, Labrador duck, Eskimo curlew, American burying beetle, striped bass[35], and the great auk.

While some of these species still exist in other regions, four of those listed above have disappeared for good from the planet: the great auk, the passenger pigeon, the Labrador duck, and the Eskimo curlew.

Never again will we hear their love songs or cries of alarm.

We are also now no longer able to obtain valuable information about their DNA composition, biocultural and acoustic heritage, and patterns of living, feeding, and reproducing.

✦

[34] The black bears of the island of Anticosti are sometimes added to the list, ever since this localized population disappeared due to intensive hunting and the introduction of white-tailed deer. Although there used to be many black bears on Anticosti, the population began to decline sharply in the first half of the twentieth century as the deer herd increased. Why did this happen? The overabundance of deer led to intensive browsing of deciduous shrubs and herbaceous plants, which then lost their ability to produce fruit. In late summer, black bears need to prepare for hibernation by feasting on wild small fruits (blueberries, bunchberries, black currents, etc.). When the quantity of these fruits falls below a certain threshold, bears can no longer maintain their body mass because they lose more energy looking for food than gained by finding it.

[35] Recently reintroduced.

The great auk (Pinguinus impennis) has been officially extinct since the mid-nineteenth century: the last pair was killed by three sailors on Eldey, a small uninhabited island off the coast of Iceland, in June 1844. For three centuries, the bird was mercilessly hunted on both sides of the North Atlantic for its meat, eggs, fat, feathers, and skin.

The great auk, known to First Nations people as *Apponatz*, was the largest of the northern hemisphere's alcids (a bird family that includes colonial seabirds like guillemots, dovekies, razorbills, and Atlantic puffins). Great auks were massive-looking, almost 70 cm tall and weighing around five kilos. They sported black plumage on their backs, white on their bellies, and large oval patches in front of each eye during the breeding season.

Although their short wings ould not support the bird's weight in flight, the great auk was a matchless underwater swimmer that was very effective in catching herring, capelin, and other fish. In this respect, it was more like penguins (family Spheniscidae), which live exclusively in the southern hemisphere.

It is believed that the great auk lost its ability to fly because of evolution, as small wings were more aerodynamically suited to streamlined bodies operating underwater.

Before Europeans came to North America, the great auk was occasionally eaten by the local population, but for them it had a more symbolic value: bone necklaces unearthed at sites dating from the Maritime Archaic period (5,500–3,200 years ago) feature representations of a large penguin; also, a costume made of several sewn penguins skins was found on top of remains in a tomb unearthed at Port au Choix, on Newfoundland's northeast coast.

This cultural fascination with the great auk goes back a long way. In Europe, 20,000-year-old cave paintings of great auks have been discovered in the Cosquer Cave in southern France.

The great auk not only lived in northern regions of America, it also inhabited the cold waters of the eastern Atlantic, notably those adjoining the islands off Scotland and Iceland, and possibly those adjoining the Faroe Islands and the Shetland archipelago.

The first documented sightings of the great auk in North America date back to summer 1534, when Jacques Cartier, an explorer from Saint-Malo, France, landed on Rochers aux Oiseaux, part of the Magdalen Islands, for a brief stopover before continuing his voyage to the interior of the continent. A passage from his logbook reflects the abundance of great auks in a mention of loading "quatre ou cinq pippes"[36] of them onto each of his two ships in less than half an hour!

A few other subsequent historical accounts have confirmed the presence of great auks in the Gulf of St. Lawrence, including that of Gabriel Sagard in 1623, who also observed the species nesting at Rochers aux Oiseaux.

The largest North American population of great auks in the sixteenth century was undoubtedly the estimated 200,000 on Funk Island, Newfoundland. However, the continuous exploitation of them by fishermen from many ships (380 in 1578 alone) accelerated the decline. The great auk was highly vulnerable because it was a poor walker and therefore nested in colonies close to the shore. From 1500 onwards, the crews of the European ships that frequented the Gulf of St. Lawrence and the Grand Banks of Newfoundland would land in the great auk colonies, club hundreds of birds to death in a matter of minutes, fill their boats, and also plunder nests for eggs.

It took just over three hundred years for a species whose population numbered several million to disappear altogether, thanks to the growth of maritime trade in the North Atlantic.

A few great auk pairs, scattered here and there on isolated islets in southwest Iceland, are thought to have survived into the first decades of the nineteenth century. One of these islets, called Geirfuglasker (Icelandic for "great auk rock"), was particularly well-suited to maintaining a small population because powerful and unpredictable currents around

[36] A quantity weighing an estimated 1,650–2,060 kg.

it prevented hunters from landing. However, an underwater volcanic eruption in 1830 swallowed up the tiny island beneath the waves.

The nearby island of Eldey was one of the last homes of great auks. However, since it was more accessible than Geirfuglasker, a series of raids led to the species' eventual disappearance when, as mentioned, the last pair were killed in 1844.

Did great auks vocalize? If so, did they grunt like their cousin, the razorbill? Or snort like guillemots? Did they make sounds in the nest or underwater when hunting? Were great auk colonies noisy? According to the few annals and historical accounts, it seems that the great auk cawed and rasped. However, given that the bird disappeared before the phonograph was invented, no great auk sound has ever been recorded.

✦

The passenger pigeon, *Ectopistes migratorius*, was sometimes called the wild pigeon or homing pigeon. Hence the possible confusion with the rock dove, a bird of European origin that was domesticated and introduced into North America in the seventeenth century.

The passenger pigeon, a native American species, was slimmer than the typical pigeon, as well as a little taller and, especially, more colourful than the mourning dove, another familiar species in our cities and suburbs.

Once upon a time, passenger pigeons were the most populous bird in North America, estimated at 3–5 billion in the mid-nineteenth century. That's right—billion, not million! It was certainly one of the world's most prolific birds.

Passenger pigeons inhabited the central and eastern United States and southeastern Canada (Manitoba, Ontario, and Quebec).

In spring, summer, and even fall, large flocks, numbering tens of thousands of individuals, visited the St. Lawrence Valley to feed on oak acorns, elm and maple samaras, and cereal grains. They nested very early, feeding mostly on beechnuts and oak acorns before these seeds germinated.

According to contemporary accounts, passenger pigeon flocks in flight were so dense and extensive that they blocked out the sun for hours on end. In 1866, people in the Niagara region watched a single flock of passenger pigeons, 13 km long, for 14 hours. It was like a river of feathers in the sky. It's also said that when the pigeons perched, tree branches snapped under their weight.

In Quebec, there are only a few records of the species' nesting. However, Pehr Kalm, a Finn, who was a student and disciple of Carl Linnaeus, the naturalist who surveyed the plants of New England and New France, provides valuable information in his diary for July 1749:

> Wild pigeons... nest in the forest. However, they don't come close to Quebec City, but rather to the forests that line both banks of the St. Lawrence a little further northeast. The ground where they nest is covered with a thick layer of droppings, up to a foot or two thick. The First Nations people never shoot these pigeons when they are brooding or have young; nor do they allow others to do so. They say it would be a serious breach of kindness towards the young because they would starve. Some Frenchmen told me that they had gone out to kill some pigeons, but the natives prevented them from doing so—first gently but then menacingly, because they could not tolerate such behaviour...

The largest colony of passenger pigeons ever observed, with an estimated population of 135 million, nested in an almost 2,200 km² area of Wisconsin.

No one could have imagined that the species would disappear one day. But that's exactly what happened. Passenger pigeons were ruthlessly hunted for commercial gain by hunters using guns and many other methods (trapping them in large nets or felling the trees on which they perched and then using sticks to beat to death any that couldn't escape). The dead birds were then sold at public markets. Their large numbers and tasty flesh made them a game bird of choice, a delicacy appreciated not only by the North American bourgeoisie, but also by small farmers and other community members. The demand for passenger pigeons became so strong that their population was decimated between 1870 and 1880.

Some American zoologists feared the worst. They acquired a few pairs and tried to breed them in captivity, but to no avail.

The last sighting of a wild passenger pigeon in the United States was in 1889, while the last such occurrence in Quebec was near the Pointe-des-Monts lighthouse on the St. Lawrence River's North Shore in May 1911.

The species became extinct in September 1914, with the death in captivity at the Cincinnati Zoo of its last representative, a female named Martha.

Did the passenger pigeon coo like other pigeons? Did it sing soft, melancholy laments like the mourning dove? Or something akin to the turtle dove's *kroo-kooooo-koo*, or the band-tailed pigeon's soft and low *hoo-whoo*?

Dense passenger pigeon flocks must certainly have been very noisy. It is said that when they settled down for the night, their chatter and arguments could be heard for miles around.

Curiously, while much is known about this extinct species, there is very little documentation about its songs and calls.

Only the oldest white oaks in America remember the lively noise of passenger pigeons roosting in their foliage.

✦

The Labrador duck, *Camptorhynchus labradorius*, is an enigmatic species. Discovered in 1789, it became extinct less than a century later in 1878, when the last specimen was seen alive in New York State. Little is known about its habitat, and nothing at all about its biology.

While the causes of its disappearance are also unclear, the small size of its population, its distribution confined to the salt waters of northeastern North America (from Labrador to Chesapeake Bay), its diet,[37] and the hunting to which it was subjected (nests were ruthlessly plundered) would undoubtedly have been the main factors in its rapid disappearance.

Labrador duck specialists think that the species nested on the rocky coasts of Labrador and Quebec's North Shore, and wintered along the New England coast. The male's breeding plumage was remarkable and gave him an aristocratic air: white wings, head, and neck; black back, belly and tail with iridescent dark green highlights; black and yellow bill; and a thin black collar at the base of the neck.

During courtship, did the male emit plaintive, *ah-OO-oo* hoots like those of the common eider? And did the female Labrador duck grunt like a king eider?

The only record of Labrador ducks in Quebec comes from La Prairie, south of Montreal. A Labrador duck was shot there during its spring migration in 1862 and its stuffed remains can be seen in the American Museum of Natural History in New York. Totally silent in its display case!

✦

The Eskimo curlew, *Numenius borealis*, is one of twenty-three species drawn by the famous wildlife painter John James Audubon during his trip to Quebec's North Shore in 1833.

In those days, the species was certainly more common than it is today. Nevertheless, there is still uncertainty about its current status because the Canadian government has not yet officially declared it extinct

[37] Probably mussels and small crustaceans: the duck's long beak, enlarged at the tip with a fleshy edge, suggests that this was a tactile organ the bird used to find invertebrates in silt and sand.

and still lists it on its Species at Risk public registry as "endangered." In fact, there has been no documented sighting with photographic evidence since a spotting in Texas in 1962, and another in Barbados in 1963. Thus, based on statements from biologists and experienced amateurs despite any formal proof for the past sixty years and more, we may very well presume that a small population of around fifty Eskimo curlews still nests, out of sight, in some remote Arctic regions.

The population of the once numerous Eskimo curlews collapsed in the late nineteenth century, due to uncontrolled commercial hunting and the disappearance of quality habitats (grasslands) during the species' spring migration from wintering grounds in Argentina to nesting grounds in the Canadian Arctic.

Hundreds of birds displayed hanging by their legs were sold for meat in Hudson's Bay Company stores or stuffed into barrels for shipment to East Coast cities. Major declines in the population were recorded from the 1870s to the 1890s, after which Eskimo curlews became few and far between.

This brownish shorebird, with its downward-curved bill, used to head back north from Argentina in spring through the centre of the North American continent. It would alight to feed on the American West's huge natural plains and Canada's extensive prairies, both of which, at the time, were already shrinking.

In fall, the bird headed for the Atlantic coast, flying over the coasts of Labrador, Newfoundland, and Quebec to build up its strength and continue its ongoing journey of several thousand kilometres over the Atlantic.

Was the Eskimo curlew as loquacious in flight as its cousin, the whimbrel? It seems so, according to the few accounts that we have. It made a loud, repeated, and monotonous *tr-tr-tr-tr-tr-tr-tr-tr* sound. That's all we know about this species' vocalizations.

Some 350 stuffed Eskimo curlew specimens are on display in natural history museums around the world. Like the great auk, the passenger pigeon, and the Labrador duck, all are silent. Forever silent.

✦

If we are not vigilant, other elements of the natural soundscape that surrounds us could disappear in the next few decades.

One candidate for this disappearance is the song of the boreal chorus frog, *Pseudacris maculata*, in the St. Lawrence Valley. Due to the destruction of this frog's native habitat in the far south of Quebec, it's not inconceivable that in the early days of spring, the distinctive *Cree-ee-ee-ee!* song of this tiny amphibian, barely longer than a thumbnail, will no longer be heard in wetlands, temporary pools, and swampy woods. The male boreal chorus frog's mating call sounds very much like the song of an insect or the rasping noise of a fingernail sliding over plastic comb teeth.

North America is the boreal chorus frog's only habitat on Earth, primarily southern Ontario and Quebec in Canada and from Tennessee to New York State in the United States. In Quebec, the frog is now found only in the Outaouais and Montérégie regions. Until the 1950s, the species was widely distributed south of the St. Lawrence River as far as the Appalachian foothills. Since then, its situation has seriously deteriorated, mainly due to urban sprawl and modern agricultural practices. These human activities have largely fragmented or destroyed the frog's habitat.

In the Montérégie region of Quebec, several populations have become locally extinct in recent decades. In 2004 alone, almost 10% of the region's breeding ponds were drained or otherwise altered. In 2008, the Committee on the Status of Endangered Wildlife in Canada (COSEWIC) estimated that the species had lost as much as 37% of its Quebec-wide population in just ten years. COSEWIC's assessment of this frog's situation is unequivocal about how severely it is threatened in Quebec. In a 2008 report, the committee noted that the habitat destruction in the suburbs of southwestern Quebec had become so rapid that the remaining populations were at risk of disappearing from their known habitats in less than twenty-five years.

In the Outaouais region of Quebec, the species is only found in a 100-km strip north of the Ottawa River. Several small, isolated populations in the greater Gatineau region have disappeared in recent decades due to residential development. And many others are even more isolated than before. In this region, the species has abandoned 30% of its original range. Since its remaining breeding habitats are in urban or agricultural areas, its future does not look bright.

✦

Up north, away from the eyes and ears of city-dwellers, the fate of the woodland caribou is of tremendous concern to both biologists and the public.

However, I won't go into the threats to these boreal forest animals in this book because I have already done that extensively in another recent book.[38]

It is chilling to think that, up there in the lichen-covered black spruce forests up north, caribou cows during the fall rutting season might never again hear the panting, bellowing, and snorting of their male suitors, nor the bleating of their fawns the following spring. It doesn't make any difference that there might not be any humans there to hear these unique expressions of biological life.

We need to be deeply worried about the quality of the soundscapes of old-growth boreal forests. The richness of this sonic heritage is a reliable indication of their state of health.

The ongoing diminishment of the soundscapes of forests, swamps, marshes, and peat bogs is a sad reflection of the degraded state of our ecosystems.

[38] *Le dernier caribou*, Éditions MultiMondes, 2020.

13

The Homogeocene

It's well-known that human population growth and its concomitant increase in resource consumption is severely impacting natural environments. The larger the population of a region or country, the more difficult it is to maintain their original ecological integrity.

Water pollution due to the discharge of excessive levels of nitrogen, inorganic chemical compounds, and drug residue is more severe in rivers flowing through densely populated areas than in rivers flowing through remote areas.

The same goes for noise pollution. With population growth, infrastructure development, and the conversion of wetlands to farmland, man-made sounds (anthropophony) are gaining ground at the expense of natural sounds (biophony and geophony).

Increased environmental noise pollution is one of the most obvious symptoms of habitat degradation not only for wild species, but also for humans and the species they have domesticated.

Five factors are disrupting ecosystems and the species they support: habitat loss and fragmentation; environmental pollution; climate change; overexploitation of natural resources, and the introduction of invasive exotic species.[39]

[39] I discuss these factors in detail in other books. Readers wishing to delve deeper may consult: *Le dernier caribou*, Éditions MultiMondes, 2020, and *Le Québec en miettes*, Orinha Média, 2012.

These phenomena also have a measurable impact on the Earth's soundscape.

Habitat fragmentation and loss

Every hour around the world, slightly more than one square kilometre (1.02 km^2) of natural surface area is lost to new infrastructure of all kinds: roads, bridges, docks, suburbs, industrial buildings, and shopping malls. This translates into exactly 1.7 hectares[40] every minute.

And that's not counting the swaths of forest cut down to make way for agriculture or to satisfy the appetite of the global forest industry. While forestry operations in most Western nations are carried out according to strict standards and sustainable resource management practices, the situation is quite different in certain parts of Southeast Asia, sub-Saharan Africa, and South America.

In Quebec, the huge areas now devoted to monocultures, mainly corn and soybeans, in the St. Lawrence lowlands, are now much quieter than the fields of yesteryear that primarily consisted of land devoted to pasture, forage crops (hay and alfalfa), and short-term fallow. These days, you can no longer hear the songs of bobolinks, upland sandpipers, eastern meadowlarks, or horned larks.

Similarly, many wetland and mature forest habitats are shrinking, and those that remain are being disconnected from their original ecosystem like islands in the middle of the ocean. It's becoming increasingly difficult for plants and animals to reach these habitats and establish colonies there, or eventually move on to settle other habitat "islands."

According to a 2019 report by the Institut de la statistique du Québec, this situation is particularly true in southern Quebec, where artificial surfaces increased by 528 km^2 from 1990 to 2013, an average increase of around 0.6% per year.

In 1994, Swedish researcher Henrik Andrén, published an eye-open-

[40] 1 ha = 100 m x 100 m (or 10,000 m^2).

ing scientific article that revealed the existence of a critical threshold or tipping point when natural habitats fragment into small isolated pieces. He concluded that if less than 30% of the original habitat area remains, the relationship between habitat area and the number of individuals of a given species on it is destabilized.[41]

In other words, considerable fragmentation effects are felt when an area loses more than 70% of its original habitat. Any species population on the remaining 30% then suffers a significant decline leading to a downward spiral towards its disappearance.

Beneath this 30% threshold, the habitat becomes so fragmented into habitat islands that it becomes difficult for species to move between viable territorial parcels. At the same time, the islands become so small that resident species struggle to maintain their existing populations. The result is an ever-worsening cascade of local disappearances of those species that are most sensitive to fragmentation. Little wonder then that these remaining shreds of nature are also becoming increasingly silent.

Habitat degradation

Pollution of water, soil, or air by various chemical products degrades habitats in insidious ways.

Even when pesticides, herbicides, detergents, heavy metals and industrial waste do not directly kill organisms, they cause long-term upheavals in living communities. One consequence is that the buzzing of different kinds of insects becomes increasingly rare in agricultural or agro-forestry landscapes.

Otherwise-beneficial minerals, like the nitrogen and phosphorus used as fertilizer, cause problems when applied at excessively high concentrations. These excessive levels stimulate algal and bacterial blooms that destabilize aquatic ecosystems and reduce the populations of in

[41] Based on a theoretical initial density of 1 individual/hectare, 100 hectares should support a population of 100 individuals; 80 hectares, 80 individuals; 50 hectares, 50 individuals, and so on.

vertebrates, fish, and all the other species (frogs, tortoises, mammals, and birds) above them in the food chain. As a result, a typical lake's once joyously noisy shores fall silent.

Atmospheric pollution generated by vehicles on busy roads also affects plant and animal species. Higher concentrations of lead were found in the tissues of insects collected near roads, compared with those collected far away from road infrastructure.

Road construction alters water drainage, leaches toxic molecules, disperses noxious particles into the air, and inevitably increases the amount of traffic-associated noise in its wake.

Willow warblers are a passerine species that typically nest in wasteland and regenerating woodland. For several years, two Dutch researchers studied the ecology of a willow warbler population living close to a freeway. They found that road traffic is detrimental to willow warbler reproduction. Specifically, they found that the species' natality was in precipitous decline in 200-metre-wide strips on either side of the freeway, despite a type of vegetation cover like that found beyond the strips. The researchers established that males did establish territories on the strips, but their reproductive success was half that of males whose territories were far from the freeway site. After multiple statistical analyses, the researchers isolated a single significant factor: noise pollution. Noise pollution from vehicles on roads can lead to imperceptible deterioration of nearby habitats.

Overexploitation of resources

Human population growth on a global scale is generating unsustainable levels of harvesting of biological resources (timber, fisheries, etc.). Every year, in the case of wild birds alone, 25 million individuals are killed (legally or illegally) by hunters with guns, clubs, or nets.

On a smaller scale, the trade (authorized or illegal) in wild species for various purposes is also responsible for the decline of many animals and plants. It is estimated that this type of exploitation affects around a third of the mammals and birds threatened with extinction. Primates (around 70,000 captured annually), reptiles (640,000 annually) and birds (around 4 million annually) are traded on the pet, zoo, and private collection markets. Aquarium fish (350 million traded annually, all species combined) are also involved. Most tropical saltwater fish that end up in pet shops are caught in the wild. Also, around 10% of orchids sold in the world are harvested in tropical regions, with many being deliberately labelled under a different identity to avoid international trade regulations.

While orchids, reptiles, and tropical fish may not be among the noisiest creatures on earth, it's hard to argue that removing four million birds from the natural habitats where they were born (not to mention the 25 million birds hunted worldwide for domestic or commercial purposes) has no effect on the world's soundscapes.

In the previous chapter, I mentioned that four bird species have completely disappeared from North America since Europeans arrived. Overexploitation is undoubtedly the primary cause of the extinction of three of them (the great auk, the passenger pigeon, and the Eskimo curlew). Three voices have thus fallen silent. How many more will do so in the coming decades for the same reasons?

Climate change

By modifying the physical and chemical conditions of natural environments, the global rise in average annual temperature directly impacts ecosystems and their inhabitants.

On the biodiversity front, climate change has the potential to upset the geographical distribution of animal, plant, and fungal species. This phenomenon has been underway for several decades. In Europe, bird communities or species groupings have already moved 37 km north, and butterfly groups have moved 114 km beyond that.

According to various studies, the northward migration of flora and fauna habitats in Quebec is taking place at an average rate of 45 km per decade, or 4.5 km per year, which is very fast indeed. Some native tree and plant species will not be able to keep up with this bioclimatic shift; for many, the average annual temperature will become too high, or the soils too dry. It is estimated that 5–20% of forest habitats could disintegrate during this century, which would lead to significant changes in forest ecosystem composition and functioning.

In temperate and boreal latitudes, populations of several insect species (beetles, butterflies, and flies) are now emerging earlier in spring than before, their shortened life cycles enabling them to adapt more quickly to changing environmental conditions. Many plants are also budding early in response to the new climatic conditions. However, this is not the case for species such as vertebrates (reptiles, birds, mammals, etc.), whose generation succession takes place over a long period of time. As a result, life cycles between prey and predator, which had taken hundreds or even thousands of years to balance out, are now suddenly out of sync. In some cases, prey has become overabundant in certain places and at certain times; in others, predators can no longer find enough food to feed themselves or their offspring during the rearing period.

The entire soundscape is bound to be affected.

Invasive exotic species

The arrival of invasive exotic plants, like Japanese knotweed, or invasive animals, like the emerald ash borer, not only diminishes the soundscapes of living communities' soundscapes, but also leads to a loss of biodiversity.

When some species suddenly find themselves outside their natural territory, some take advantage of the situation to settle into their new home with too much success. The increasingly homogenous global distribution of many species that are colonizing new worlds is such that it is now common to speak of the advent of a new geological era: the Homogeocene.[42]

It's as if we'd gone back 200 million years, to the original Pangea, that super-continent made up of all the continents fused together, before it cracked and allowed a unique natural world to emerge on each of its parts. While species have been imported and exported to the four corners of the globe since the Renaissance, the recent acceleration of global commerce means that the movement of plants and animals, whether voluntary or involuntary, is now more intensive than ever. One reason for the ecological success of species in new habitats is the absence of the predators, competitors, or parasites that usually control their population growth. Without enemies, these species proliferate to the point of profoundly altering the plant and animal communities that host them, and consequently the soundscape.

A typical example of the biophonic modification of a natural North American environment by an invasive exotic species could be a marsh whose perimeter has been invaded by the native Eurasian strain of the common reed (*Phragmites australis*). Before this species settled on the marsh perimeter, the marsh's diversity of aquatic plants and wetland vegetation attracted invertebrates, frogs, birds, and mammals. Yet, once the invader takes over more and more territory and begins ex-

[42] I prefer the concept of the Homogeocene to that of the Anthropocene (frequently used to designate the current geological period, i.e. the one now scarred by the effects of human activities on the Earth). In my opinion, the Anthropocene concept is too closely associated with an anthropocentric vision of the world, as if everything revolved around us humans. More importantly, this "us" ignores the fact that the impact of Homo sapiens is the historical result of the role played by a small fringe group in human society. Climate change, habitat loss, pollution, etc. are the consequences of a model of natural resource exploitation (mines, forests, fisheries, etc.) applied by Western industrial societies. This model is based on inequitable processes involving cheap labour, even slavery, as well as the usurpation of all the ancestral territories of the planet's indigenous peoples.

tracting water and nutrients from the soil more efficiently than native species, it ends up dominating the whole area. As the lone competitor, it completely simplifies soil structure (only one vertical vegetation stratum left) and food diversity (*zero* flowers and fruit produced by other plants). The animal community in terms of birds alone is reduced from dozens of species (herons, shorebirds, and several warbler and bunting species) to a small, limited group of generalist birds with few requirements in terms of habitat and ecological conditions. Only species that are abundant in many types of unforgiving environments, such as the red-winged blackbird or the song sparrow, flourish. The result, once again, is an impoverished soundscape: where once a symphony orchestra played all that now remains is a string quartet...

Which often plays only a single score.

✦

The combined effect of the five factors mentioned above (habitat loss and fragmentation; environmental pollution; climate change; overexploitation of natural resources; and the introduction of invasive exotic species) can not only simplify, but also homogenize natural ecosystems. Climate change acts in synergy with the other factors. In southern Quebec, landscape fragmentation, accompanied by higher temperatures, is conducive to the spread of not only invasive exotic species, but also disease vectors like West Nile virus and Lyme disease.[43] Another example is the northern reproduction boundary for that well-known invasive exotic shrub, Japanese knotweed. This is now located in the Quebec City region, 500 km north of its former limit around Boston, Massachusetts, at the beginning of the twenty-first century.

It would be safe to say that it will be the already common and wide-

[43] The white-footed mouse, the primary host of the blacklegged tick, the species responsible for the spread of Lyme disease, is increasingly found in southern Quebec. This mouse is considered a very effective transmission agent in southern Quebec's fragmented and now warmer habitats.

ly spread generalist and opportunistic species that will benefit from these ecological upheavals to the detriment of those species that are less abundant and more specialized in their search for food and breeding habitat.

Biodiversity loss mainly affects fields, pastures, forest plantations, and urban environments, i.e., habitats whose dynamics are controlled by humans. In these environments, the decline in biodiversity (number of species) and in populations of the remaining species (number of individuals of each species) can reach 30–50%.

Tomorrow, ubiquitous, generalist species will find themselves in a monotonous world, with the same plant and animal communities from one environment to the next: a world where European buckthorn (another invasive shrub from elsewhere now establishing itself in North America) will dominate regenerating woods, while beds of garlic mustard (an invasive exotic herbaceous plant from Europe) will rule the undergrowth. A world where the crow will be queen and the American robin the king. A world where the spring dawn concert of all the other passerines that once contributed to the acoustic richness of their habitat diminishes day by day.

Welcome to the Homogeocene.

14

Quartet for the End of Time[44]

Through trial and error, countless detours, and the interplay of evolution and natural selection, living, murmuring nature has constantly renewed itself over the course of geological time. Despite planetary disruptions such as ice ages, widespread volcanic eruptions, large meteorites hitting the Earth, and mass extinctions, life has always resumed its course, diversified itself, and invented a thousand new forms and functions, and a thousand new songs.

That being said, conservation biologists make no bones about it: we have well and truly entered a phase of accelerated biodiversity decline. Unlike a sudden catastrophe, this decline is difficult to discern because it is spread out over time and will continue for many decades and even hundreds of years. This situation is real, and it is profoundly alarming.

Just as visual landscapes are gradually being transformed in our environment, natural soundscapes are being eroded by a combination of almost-invisible changes. Year after year, in our countrysides, towns, and cities, we are slowly getting used to hearing fewer insects and fewer songbirds like swallows and swifts.

This means that the natural soundscapes of the future will be more similar to string quartets than symphony orchestras.

[44] This chapter title is taken from *Quartet for the End of Time*, a musical work for violin, cello, piano, and clarinet composed by Olivier Messiaen and inspired by the Apocalypse of St. John. Messiaen composed the work while a prison camp inmate in Germany in 1941.

Imagine an avian quartet consisting of a a cawing crow, a chirping sparrow, a cooing pigeon, and a humming blackbird. And then, another made up of a cheekily sniggering ring-billed gull, a disconcertingly hiccupping red-winged blackbird, a brightly chattering song sparrow, and a discordantly squawking common grackle.

In its own furtive way, the current biodiversity crisis marks a truly large-scale extinction—the sixth to occur since biological life first appeared on the planet. This crisis has the hallmarks of the five previous mass extinctions as defined in the scientific literature:

1. It is global, or at least broadly based;
2. It is brutal and long-lasting, and may extend on a geological scale over as many as a million years;[45]
3. It simultaneously affects many different species groups, from molluscs, worms and crustaceans in the sea to vertebrates (birds, reptiles, and mammals) on land.

So, let's rewind the film of life on the planet to measure the scale of previous crises and their consequences on past soundscapes.

Some 445–440 million years ago, life mainly existed in the oceans. Marine biodiversity consisted of no fewer than 500 families—sponges, bryozoans, trilobites, molluscs, and jawless fish. In sheltered reefs of shallow, warm waters, clams and anemones filtered a diet of rich organic particles borne along on currents, and large snails munched on algae. Strange toothless and finless fish, covered in thick protective plates, patrolled the ocean bed.

This world was largely dominated by geological sounds: waves crashing against natural breakers, falling rain, and winds sweeping over emerging lands. The only sounds generated by animals were incidental and accidental: the clattering of mollusc shells as they abruptly closed or the sound of fish bellies' rubbing along the ocean floor. Then came a short, brutal and intense ice age known as the Ordovician. As a result

[45] One million years is a relatively short period in the 3,800 million years of life on Earth.

of abrupt climate change and a rapid lowering of sea levels, 60–70% of the biological species on Earth were wiped out.

Life, still primarily based in the oceans, took time to recover. A few million years later, the oceans were once again teeming with a huge variety of species: starfish, worms, crustaceans, and other arthropods roamed the mudflats and other ocean bottoms in the vicinity of reefs; fish now had jaws; and the first sharks roamed the open waters. Then came another mass extinction 370–360 million years ago, this time caused by an abrupt drop in oxygen levels in the water—the Devonian. Once again, sudden environmental changes like another radical climate change or abrupt fluctuations in sea levels or even a meteorite's impact are thought to have triggered this mass extinction of almost 75% of all species. If a meteorite was indeed the cause, it's hard to imagine the explosion that would have accompanied the fall of such a huge object from outer space. It would probably have been the loudest sound ever heard on our planet's surface. If any of the surviving animals had a sense of hearing, they would have been immediately deafened.

Life continued on a new course, until another, even more dramatic crisis—the Permian extinction about 250 million years ago. This event, the most damaging such event in our planet's known history, devastated life both in the water and on land. Many geologists still suspect that an asteroid impact was the cause; other scientists are more inclined to think that intense volcanic activity was the cause, as evidenced by recently uncovered traces of large-scale basaltic lava flows that covered Siberia at the time. Such intense volcanic activity would have spewed out an unprecedented volume of particles and greenhouse gases into the atmosphere, precipitating rapid global warming on a geological scale. The average temperature of the planet would have risen from 12–15°C to 22°C, a jump of almost 10°C! Around 95% of living species disappeared during this period with virtually every group affected—fish, amphibians, reptiles, herbivores, carnivores, plants, insects, and other arthropods. The Permian extinction has been by far the most destructive of all the Earth's five mass extinctions to date. Yet, we know that at that time, some animal species, notably crickets, already possessed the ability to hear and make sounds. However, other animals that had begun to vocalize were either eradicated forever, or became mute.

After all that devastation, it took a while for life to return. But once again it appeared in new forms. Vast evergreen rainforests sprang up, with a plethora of animal species livening up the soundscape with cries, songs, and reverberating grunts, while their movements rustled the foliage. Giant bush-crickets of a now extinct suborder called Titanopterans courted each other with guttural, low-pitched sounds reminiscent of bullfrogs, while the dawn resonated with insect stridulations, awaiting the melodious tunes of songbirds that would not be heard for tens of million years more.

Then, just 50 million years or so after the Permian came another extinction: the Triassic around 210–200 million years ago. This event once again wiped out almost all life on Earth: some 75% of all species, including archosaurs (dinosaur ancestors) suffered the wrath of fate with the loss of their skin, feathers, or scales. Several hypotheses have been put forward to explain this catastrophe, including another episode of extreme volcanism, followed by another global warming. According to some researchers, the fall of a huge meteorite five kilometres in diameter 214 million years ago in Quebec,[46] may have exacerbated this crisis.

The most recent and best-known mass extinction is the Cretaceous, around 65 million years ago. Prior to that, dinosaurs and other reptile families with diverse forms and functions, prospered, as did birds, mammals, amphibians, and fish. At the same time, insects and flowering plants thrived symbiotically, the latter offering their nectar to the former, and the former, in turn, fertilizing plants by spreading their pollen. Biodiversity abounded, and soundscapes grew richer and more complex. Dinosaurs, albeit not particularly noisy creatures, hissed and

[46] The impact site is known as the Manicouagan Reservoir on Quebec's North Shore.

grunted, frogs sang, and small crepuscular mammals babbled discreetly in thickets.

Then another end-of-the-world event led to the upheaval of food chains both on land and in the sea. As with previous mass extinctions, speculation continues as to the cause of the Cretacean, the fifth major eradication of life on Earth, with the two most common hypotheses being another period of extreme volcanism or the fall of a meteorite in Mexico.

Subsequent recovery this time was laborious, with plants particularly slow to re-establish themselves and diversify. Over a period of several million years, vast ecosystems hollowed out by the disappearance of the dinosaurs were slowly recolonized by several hundred families of mammals, birds, and other living beings.

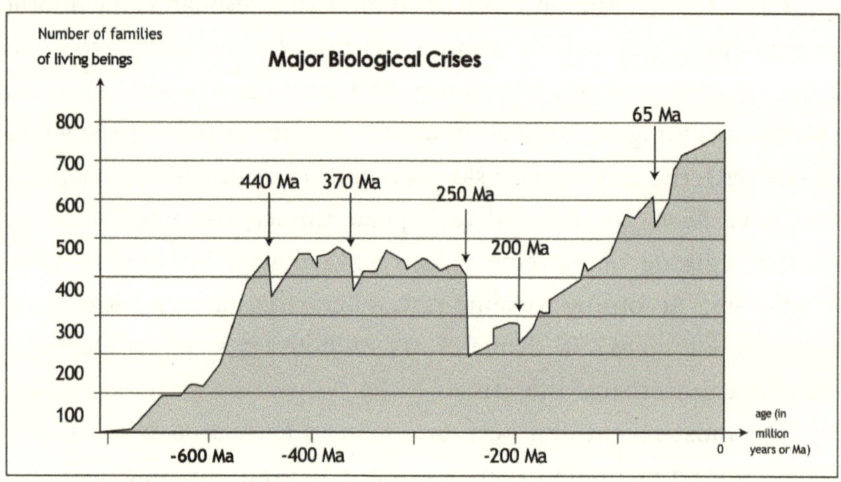

Table 6 The five greatest extinctions since the appearance of life on Earth, occurring in the Ordovician (-440 Ma), Devonian (-370 Ma), Permian (-250 Ma), Triassic (-200 Ma), and Cretaceous (-65 Ma) periods.

Around 15 million years after the disappearance of the dinosaurs, mammals had already diversified to the point of 75 different genera, divided into numerous orders. These mammals were not only agile, intelligent, and endowed with tremendous behavioural flexibility, they could also produce and decode a host of signals and sound stimuli to take advantage of the new world opening up to them. The mammalian evolutionary tree sprouted numerous highly developed branches, most notably bats, flying mammals, and primates with enormous brains. The latter group gave rise to a species, *Homo sapiens*, which could single-handedly be held responsible for the next mass extinction—the one we could very well be witnessing.

✦

Some researchers, however, refuse to believe in this gloomy fate, arguing that the rate of species extinction estimated by proponents of the sixth extinction thesis is exaggerated. Based on the International Union for Conservation of Nature's IUCN Red List of Threatened Species™, they argue that very few species have become extinct since the explosive increase in the human population since the late Middle Ages. The extinction rate calculated on the basis of the Red List would be less than 1%, i.e., a normal rate of species extinction, given that terrestrial species naturally disappear over time at a constant, predictable rate.

However, vertebrates (mammals, birds, reptiles, amphibians, and fish) are over-represented on this list, which skews the situation significantly. In 2022, Robert Cowie of the University of Hawaii, assisted by two French researchers, Philippe Bouchet and Benoît Fontaine, used other species groups, such as terrestrial molluscs (slugs and snails) to show that the actual extinction rate was significantly underestimated. They calculated that 7.5–13% of terrestrial mollusc species had disappeared since 1500. Thus, if these proportions were applied to all living organisms, 150,000–260,000 living species might very well have become extinct since the end of the Middle Ages. For the most part, these species, like many insects, worms, and other invertebrates, would be discreet, anonymous, unclassified, and invisible to our eyes, but nonetheless essential to ecosystemic balance. That is why we are witnessing the beginning of the sixth mass extinction. To deny this or to accept it without doing anything is morally indefensible and will only precipitate the planet more rapidly into the anticipated catastrophe.

On the other hand, we should not forget that, as in the aftermath of previous mass extinctions, life always resumes its course, sometimes in surprising, innovative, and roundabout ways.

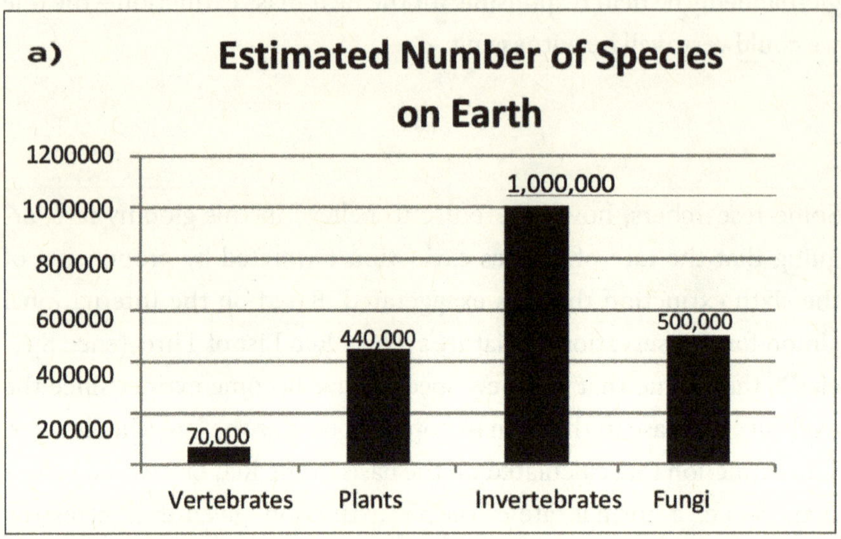

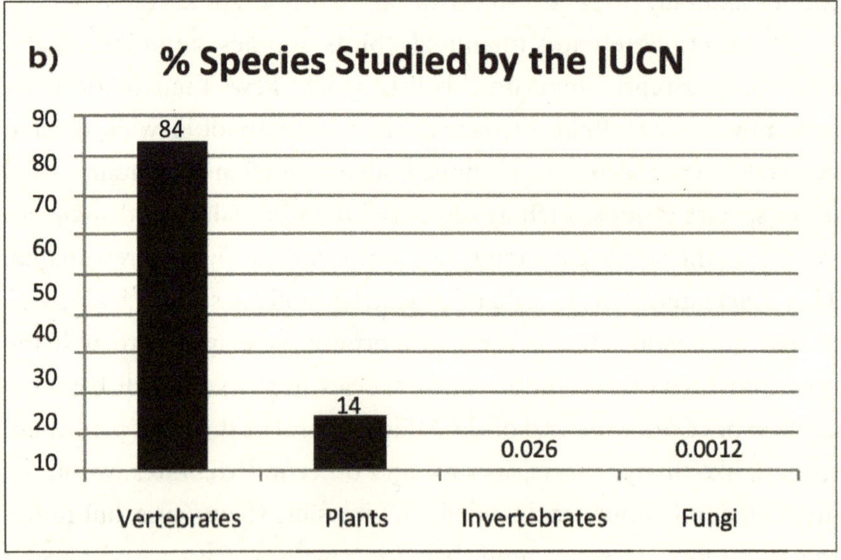

Table 7

There are no reliable tools for measuring species loss on a global scale. The current rate of species extinction is therefore unknown. This is particularly true for invertebrates (insects, spiders, molluscs, etc.), which account for an estimated 1 million species [see Table 7 (a) above]. On the IUCN Red List of Threatened Species, vertebrates (mammals, birds, reptiles, amphibians, and fish) and vascular plants are over-represented, compared with lesser-known species groups such as invertebrates and fungi [see Table 7 (b) above]. Only 0.026% of invertebrate species populations and 0.0012% of fungal species populations have been studied by the IUCN. (Adapted from Cowie et al., 2022)

It is therefore safe to say that the living world will certainly recover from the crisis for which we are currently responsible. The harmony and music of that new world will certainly be different. It's quite possible that one of the byways life will take to be reborn will involve a virtual extinction of our own voice on Earth.

✦

Natural ecosystems at the local level sometimes irreversibly shift from one state to another when they cross one or more critical thresholds. An example of a critical threshold could be that of a limpid lake, ideal for swimming, which is invaded by blue-green algae (cyanobacteria) due to a sudden increase in the phosphorus and nitrogen in the lake. The presence of these organisms rapidly alters the lake's ecological balance. Not only does the lake suddenly become unfit for swimming, but large quantities of algae block fish gills, modify the amount of dissolved oxygen in the water, and accelerate the aging of the body of water (eutrophication). Once this process has begun, it is very difficult to "rejuvenate" such a lake by returning it to its original state.

Is it possible that the biosphere—the global ecosystem—could react in the same way? When it goes beyond a critical threshold on a planetary scale, could the Earth descend into an irreversible and irrevocable catastrophe?

That's the view of a group of twenty-two scientists from around the world who published their thesis in the scientific journal *Nature* in 2012. In their view, the planet is approaching such a tipping point due to human activities.

In terms of climate change alone, any further increase in the planet's average temperature over the next few decades, no matter how slight, is likely to lead to irreversible forest dieback, the extinction of marine life in coral reefs, the permanent thawing of permafrost, and so on. In 2022, a team of ten researchers led by David Armstrong McKay of the University of Exeter in the UK, wrote in *Science*, another highly reputable scientific journal, about just such a climate tipping point.

✦

As discussed in previous chapters, the biosphere is being battered by factors such as human population growth, intensive, consumption-driven extraction of natural resources, habitat fragmentation and degradation, and climate change. The consequences for natural environments far exceed their capacity to regenerate.

According to the Intergovernmental Science-Policy Platform on Biodiversity and Ecosystem Services, 43% of the Earth's land surface has been converted to predominantly agricultural or urban land. These lands now host only a few plant crops (e.g., corn, soybeans, wheat, and rice) and animal groups (e.g., cattle, pork, and poultry), all of which are produced for human consumption. Similarly, fishery resources are diminishing year after year due to fishing, and ocean waters are becoming more acidic due to rising CO_2 concentrations.

Other effects are broader. For example, human activities are profoundly changing how energy circulates in the biosphere. Human activity now monopolizes 20–40% of the world's net primary productivity represented by the photosynthetic chemical energy produced by plants on the planet's surface after subtracting the portion needed to maintain the plants themselves. This surplus biomass produced by plants normally becomes available to all other living organisms (animals, bacteria fungi, etc.) that cannot photosynthesize. However, humanity is now diverting a large part of this surplus biomass for its own benefit.

On continents and oceans alike, many species of plants, vertebrates, invertebrates, and fungi now exist within narrower habitat ranges, and their once-abundant numbers are sharply declining. Although these organisms are not currently included on national lists of extinct species, population figures for several classes of living organisms—mammals, birds, amphibians, insects, flowering plants, etc.—nevertheless indicate that we are approaching a critical point. Increasingly depleted and less stable ecological communities do not bode well for the future.

Feedback loops between different phenomena could rapidly knock many ecosystems off balance, plunging them into chaotic states. In the space of just a few years and on a smaller and more local scale, rapid climate change, combined with the invasion of exotic species and a severe contraction in the range of native species, is able to turn rich, productive marshes populated by dozens of birds, mammals and butterflies into unrecognizable wet wasteland, dominated by a handful of foreign plants, such as Japanese knotweed or common reed that host a few insect species, sometimes exotic like the Japanese beetle, or a few opportunistic bird species like the red-winged blackbird.

Is such a change of state plausible on a biospheric scale? Researchers who have looked into this question say yes. Apart from being less diverse and more homogenous, such a natural world would also be more unstable and chaotic.

But it's possible that those predictions won't come true. That is why we should take hope and get to work.

✦

The song of the crickets...

Intentional animal sound communications were first heard on our planet in the forests of Pangea during the Permian period some 272–260 million years ago. It was a genuine transmission between a sender (a male cricket) and a receiver (a female cricket). The first call, the first message of love.

Over the course of evolution, other orthopteran insects—bush-crickets, grasshoppers, and the like—also adopted mating calls, as did the ancestors of amphibians, birds, and mammals.

Whether at dusk, in the dead of night, or just before dawn, the males and females of tens of thousands of animal species still call out to each other. Many of these ardent choirs' calls are about to disappear if we're not careful.

15

A New Score for the Earth's Soundscape

The impact of a single species—ours—is currently plunging the planet into a biological crisis equivalent to a powerful tsunami ravaging the shores of a magnificent country. We are rapidly depleting a natural capital that took 3.8 billion years to build up.

We don't realize that this is happening because our perception of time, our reference time scale, is not the same as that of the rest of the living world. Every second hundreds of stocks and shares are exchanged around the world, every month people pay their mortgage or rent, and every four years or so, politicians, at least in Canada, are re-elected.

The time scale of living organisms, on the other hand, is much broader, ranging from a few decades (a few generations on our human scale) for a forest to grow again after a catastrophe to a millennium for a drop of water to be carried around the globe by ocean currents. Geological time is quite a different story—it is counted in millions of years.

What we do realize, however, is that ecosystem health and human health are closely linked. One cannot exist without the other. When an ecosystem is degraded, it can no longer deliver its ecological benefits or services to humans with its usual efficiency. Such important services include atmospheric carbon capture, air and water filtration, flood mitigation, groundwater replenishment, and fruit and vegetable pollination by insects.

To solve the biodiversity crisis, we first need to wake up to it, and accept the facts and the alarm bells coming from nature itself.

The profound modification of the planet's soundscapes is undoubtedly one of the most obvious symptoms of our ecosystems' failing health. But the good news is that it's not too late.

✦

To protect biological diversity, the solution is the same as that to protect natural soundscapes: we need to remove much of the land and the sea from human influence.

It's apparent that only very few ecosystems, such as deep ocean trenches, certain remote parts of the Amazon rainforest (for how much longer?) and the taiga and arctic tundra closer to home, remain largely unaffected by human activity. Everywhere else on the planet except for Antarctica, humans have established themselves or are exploiting natural resources.

Mother Nature's tolerance threshold has been accurately measured at around 30% of natural surface area. Below this limit, a given territory's biodiversity can be seriously altered and the integrity of its local environment cannot be maintained. And, as mentioned, humans might very well no longer benefit from the ecological services that biodiversity provides.

Conservation specialists have known of the existence of this threshold for over thirty years. A recent study, published in the summer of 2021 by researcher Martin Jung and 51 other colleagues from various universities and institutions around the world, reconfirmed its validity. He and his colleagues pointed out that staying below this threshold would also significantly limit the impact of climate change, especially by ensuring groundwater replenishment to guarantee freshwater availability.

The 30% target for protected natural environments was also on the agenda for COP15, the 15th Conference of the Parties to the United Nations Convention on Biological Diversity, held in Montreal in 2022 (after being transferred from Kunming, China). At the conference, the 196 signatory nations of the Convention on Biological Diversity (CBD) focused on achieving 22 other targets by 2030. These included a major reduction in the use of pesticides in agriculture, an end to overfishing in the oceans, and intensified efforts to combat invasive species.

In his opening remarks to the conference, United Nations Secretary-General António Guterres warned the 18,000 delegates in attendance that humanity itself had become "a weapon of mass extinction." This rather blunt statement for such an experienced international diplomat set the tone for negotiations by highlighting the urgency of the situation.

Elisabeth Maruma Mrema, the CBD Secretariat's Executive Secretary, issued a similarly stark warning when she said: "We are no longer approaching the point of no return, we are here." She emphasized the need for ambitious and immediate action and that it was no longer possible to procrastinate.

Fortunately, in the final hour of the conference in the middle of the night, the delegates adopted by consensus the Kunming-Montreal Global Biodiversity Framework (GBF), put forward by China, which was presiding the session. According to the delegates' closing statement, the GBF represents "peace with nature." Its implementation now depends on every signatory country.

Will we succeed? Let's hope so. Because what's the point of developing an ecological science capable of making predictions, only to do nothing in the end?

Imagine coming across a message in a fortune cookie announcing the time of your own death. If it told you that you only had a few years to live, what would you do? If your fate were inescapable (you're going to die in a road accident or a plane crash), you'd probably choose to carry out a project close to your heart, travel, do something memorable, or settle a dispute with a loved one. But if your fate were reversible, for example, an illness linked to bad lifestyle habits (such as smoking or being sedentary), you'd probably decide to get in shape or stop smoking to postpone your death for as long as possible.

Ecological science has written the almost exact moment of the biosphere's, and our civilization's, tipping point on our imagined fortune cookie message. Although it's not an irreversible fate, we're acting as if nothing is written on that piece of paper inside the cookie. We simply don't want to read it; we don't want to know.

We lived carefree in the twentieth century, until the 1960s, when Rachel Carson's book *Silent Spring* made it clear that our soaring populations and technological prowess were having considerable impact on nature and our living environment.

But we carried on as if it were business as usual—mowing the lawn, spraying pesticides, and dusting the house, even though it was in flames.

✦

Putting land in protected reserves and thereby forfeiting access to its natural resources is a direct challenge to the economic foundations of our societies. Limiting development and the sacred cow of economic growth? Easier said than done. Wouldn't a general mobilization to protect biodiversity be unrealistically utopian? But is that really true?

Take, for example, the Montreal Protocol, an international agreement between 197 countries signed in 1987. In record time, this agreement put an end to the use of industrial chlorofluorocarbons (CFCs) that had depleted the atmosphere's ozone layer that protects us from the sun's ultraviolet rays. Since the Protocol was signed, other products less harmful to the ozone layer have been substituted, and CFCs are no longer being used. As a result, the hole in the ozone layer is shrinking and the layer itself is recovering. It is predicted that the ozone layer will be completely restored within fifty years. Isn't that proof that, as a species, we can react quickly and deal with a problem, no matter how big or complex?

Another success has been the fact that COP15 ended on a note of optimism: the 30% target for protecting natural areas on land and sea was declared absolutely essential even though agreement was not achieved on several other quantified targets. In other words, a bittersweet, mixed outcome in the fight against biodiversity loss.

These days, more people are taking the protection of nature very seriously. They understand what is at stake and are prepared to make sacrifices to achieve this, such as eating less meat, controlling the consumption of goods, and paying more for sustainably harvested lumber.

On the other hand, some of us are still dragging our feet because our lives and profits directly depend on exploiting natural resources (think of certain players in the forestry industry) or pursuing real estate projects (think of certain land development companies). Not to mention all those of us who have already thrown in the towel, arguing that it's already too late, that the game is already lost, that humanity itself is doomed to failure or, at the very least, to becoming bogged down in increasingly harsh living conditions for hundreds of years.

In a context of limited financial resources, conflicting uses of coveted territories, and the scarcity of available land to be put under complete protection, two criteria can be used to prioritize the sites to be included in the goal of 30% protected areas: (1) their distinctive character and (2) their vulnerability.

A given territory's distinctive plant and animal communities, i.e., those under-represented on a regional scale, should be the first to be protected. In southern Quebec, for example, old-growth hardwood forests are home to increasingly rare forest communities in the St. Lawrence Valley. These forests provide shelter and food for precious bird species like the wood thrush, the Eastern wood peewee, and the cerulean warbler. Who would want to live in a world where it's no longer possible to hear one of the most beautiful songs in North America, that of the wood thrush? Who would want to walk in a forest where the muted whispers of flying squirrels have disappeared? Or stroll in a woodland deserted by its bats?[47]

The second criterion of vulnerability recognizes that some ecosystems, due to their ecological processes, their resident living communities, their small size, or simply their proximity to an urban, periurban or agricultural environment, are more vulnerable than others to being disturbed by pollution, habitat fragmentation, invasive exotic species, and so on. Maintaining these ecosystems within an adequate network of protected areas means preserving the full range of life in each region in all its biological and ecological diversity.

We can achieve our goal with the help of a few measures: an accurate picture of the different natural environments of regions or countries; sufficient surface area for environmental redundancy (several replicate or similar sites to offset any serious problem at a particular site); and system resiliency (the ability of the component ecosystems to return to their initial state after a disaster or other major disturbance).

✦

While protecting large wilderness areas in Quebec's subarctic and arctic regions, for example, is a realistic objective, the same exercise in southern Quebec's urban, forestry, or agricultural areas would be a huge undertaking.

In December 2020, Quebec achieved the 17% target for protected areas on its territory as set by the international community for 2020.[48]

[47] Some bat species are sensitive to noise generated by human technology, especially when vehicles or other equipment emit ultrasound of the same wavelengths as those that bats use in their search for prey.

[48] The year 2020 was a pivotal one for the international community, since during the International Year of Biodiversity in 2010, the vast majority of countries had pledged to achieve 17% protected land areas and 10% protected coastal and marine areas in their respective territories with the next decade.

In just a few years, the province went from being bottom of the class in Canada to being a model student. The very next year, Quebec, as proposed by the CBD, committed to protecting 30% of its natural environments by 2030. So much the better.

However, most of Quebec's currently protected areas lie north of the 49th parallel. This means that we're still a long way from having enough area under conservation in southern Quebec, as well as a fair ecological representativeness of the south's living communities. Of course, given that most of the land in the south is privately owned, it's more difficult to protect vast areas belonging to a single owner. Still, since it's mainly in the south that the two criteria of distinctiveness and vulnerability apply, it's south of the 49th parallel where we need to redouble our efforts.

Southern Quebec's biological diversity is irreplaceable, yet vulnerable.

✦

To achieve the CBD's conservation targets, it will be necessary to approach accessible areas in terms of land use.

It's delusional to think that we can conserve 30% of the city of Montreal's territory, for example. On the other hand, we could do so in the Montreal metropolitan area (Communauté métropolitaine de Montréal) that encompasses the 82 municipalities of the northern and southern rings surrounding the central island by making up the shortfall on the Island of Montreal with a higher proportion of up to 40–50% of protected land in the other municipalities.

A working group of the IUCN's World Commission on Protected Areas recently examined realistic targets for conserving biodiversity and reducing global biodiversity loss on a broad spatial scale. The group's researchers identified three main landscape types in a way that integrates biodiversity-related factors and pressures, both nature-centered (what's left of it) and human-centred (land and resource use). These three landscape types are: urban and agricultural (towns, cities, and farms); mixed (shared lands); and pristine (vast wilderness areas).

If implemented, this approach would facilitate action on the scale required, while ensuring that each country has a practical strategy to remedy its particular problems in keeping with global standards that are fair throughout the world.

To improve biodiversity for each landscape type, a set of solutions and best production and natural resource use practices has been proposed. Let's take a closer look.

Urban and agricultural landscapes: Primarily consisting of towns, cities, villages, and farms, these landscapes cover 18% of the Earth's surface and are home to 75% of the planet's human population. Only 6% of these areas are protected in any way. The goals of the main lines of action proposed by the IUCN working group for these landscapes are to protect threatened species and remaining ecosystem fragments, restoring them where necessary; integrate sustainable practices such as protecting productive agricultural land and practising regenerative agriculture, including the fostering of healthy populations of pollinating insects; and plan greener cities to reduce carbon emissions and slow down urban sprawl. Although it's not realistic to aim for 30% habitat protection in these areas because natural environments are few and far between, achieving 17% protected natural environments at the very least would maintain the ecological representativeness of these lands, while guaranteeing access to nature for the sake of city dwellers' health and well-being.

Mixed landscapes: Characterized by land divided between nature and natural resource production or extraction (mining, forestry, etc.), these landscapes cover 56% of the Earth's surface and are home to 25% of its human population. Around 14% of this land is protected. The main lines of action proposed by the working group for these landscapes are to establish protected areas that are ecologically representative and sufficiently interconnected; restore ecological processes and native species populations, where necessary, and otherwise maintain them; implement sustainable practices and processes for natural resource extraction and other activities, such as ecotourism, hunting, and fishing; integrate Indigenous knowledge into the protected area networks; and aim for 33% protected area per major ecological region so as to maintain adequate geographical representativeness of ecosystems.

Pristine landscapes: Primarily consisting of large wilderness areas, these landscapes cover 26% of the Earth's surface but are home to only 0.1% of the human population with some 24% of their surface area protected. The goals of the main lines of action proposed by the working group for these landscapes are to conserve overall ecological integrity and associated processes, such as carbon storage, river flows, major animal migrations, and water and precipitation cycles; prevent further fragmentation by allowing only a few intense industrial development nodes in regions that are primarily wilderness; remove or rehabilitate as necessary certain sites in Quebec such as abandoned mines; support traditional conservation management practices based on Indigenous knowledge in order to establish large protected areas and Indigenous zones to be maintained by local communities; and target 50% protected areas per eco-region in order to maintain these landscapes' ecological and geographical representativeness.

If the proposed protected area percentages for the three landscape types (17% + 33% + 50%) are added up, we get a total of 100%, or a 33% protected area figure overall for a given province or country. Bingo—we will have achieved the scientists' recommended target!

If the approach of recognizing landscape types and setting different targets for the percentage of the area of each type to be protected were applied to Quebec's territory, we would end up with three rough targets.

One target would be to maintain 17–30% protected areas in the major municipalities, their suburbs, and the outlying densely populated agricultural areas of the St. Lawrence Valley around the 45th and 46th parallels.

Another target would be 30–50% protected areas in less populated agro-forestry and rural landscapes, such as Témiscamingue, Saguenay-Lac-Saint-Jean, and Bas-Saint-Laurent in the Laurentian and Appalachian foothills between the 46th and 49th parallels.

The third and last target would be 50% or more protected areas in the boreal and arctic wildernesses above the 49th parallel.

✦

A possible solution for making rapid progress south of the 49th parallel would be to transform the wildlife reserves operated by the Société des établissements de plein air du Québec (Sépaq), the Quebec government's agency for parks and wildlife reserves, into protected areas.[49] By declaring southern Quebec's twelve existing wildlife reserves as protected areas, we would add almost 35,000 km² to the conservation bank of Quebec's natural spaces. A prodigious leap forward for the south! If the Réserve faunique de Port-Cartier–Sept-Îles, a wildlife reserve covering 6,523 km² between the 50th and 51st parallels on Quebec's North Shore (currently being considered for designation as a Quebec national park) were added to the mix, it would give us a total of just over 41,500 km² of protected areas[49] in southern Quebec.

By the same token, permissible public uses on these lands would have to be modified to safeguard ecological integrity, with responsibili-

[49] This idea was suggested nearly fifteen years ago in my book *Nous n'irons plus au bois* (Vélo Québec, 2010).

ty for these lands' natural resources being transferred to Sépaq.[50]

It is possible to hunt, fish, and log in Quebec's wildlife reserves even though the IUCN considers that logging or any other industrial activity involving the harvesting of a natural resource is incompatible with a given area's eligibility for protected status. At the same time, the IUCN accepts that regulated hunting in a designated reserve is a compatible activity. In this respect, when hunting and fishing are carried out within acceptable limits for population renewal, these activities comply with international requirements for certain types of conservation areas.

That said, Sépaq is not responsible for the planning and use of the natural resources on its territory, and, in fact, has no say when a logging or mining company operates on its territory.

If logging or mining operations were out of the picture, new protected areas could be given special conservation status, simply by adding hunting (it's already possible to fish in all of Quebec's national parks), so long as the activity is regulated and is compatible with the ability of harvested species to maintain their populations.

Other areas, such as existing natural environments that provide ecological connectivity between large protected conservation cores could also form part of southern Quebec's protected areas network. By granting a less restrictive status than that assigned to national parks or formal protected areas, these connectivity zones would authorize certain uses, such as ecotourism, low-density vacation resorts, certain other outdoor activities, and the traditional and sustainable exploitation and use of certain natural resources, while also ensuring that flora could migrate and fauna could move freely between the large protected territories. Although Quebec still does not recognize such effective conservation measures, other Canadian provinces like British Colum-

50 The management and use of natural resources (forests, mines, oil, gas, and hydroelectric or wind power) in wildlife reserve areas are currently governed by the Quebec Ministry of Natural Resources and Forestry.

bia use them as an integral part of an effective protected area network.[51]

✦

In recent years, researchers and conservation specialists have organized symposia, consultations, and working groups to look at how to achieve realistic biodiversity protection in Quebec.

Some promising ideas have emerged, such as the creation of a commission for the protection of natural land along the lines of the province's commission for the protection of agricultural land—the Commission de protection du territoire agricole du Québec (CPTAQ), created in 1978.

The CPTAQ's relevance is unquestionable, given that its mission is to protect agricultural land from urban development pressures. According to the Commission's mandate, land in an agricultural zone may not be divided up, altered, or used for purposes other than agriculture, except with special permission and a certificate to that effect.

Although some agricultural land has been removed from the zone for non-farming purposes, the CPTAQ has, thanks to its supervisory role, generally been able to conserve thousands of hectares of arable land in the St. Lawrence Valley.

The creation of a similar commission to protect natural land would make it possible, on a regional basis, to protect environments of high ecological value, identified as priorities or at least important for ecological connectivity. Such a commission could be mandated to achieve protected area percentages for each of the three landscape types.

The fact that many aspects of land management in Quebec, have been delegated to regional county municipalities (RCMs) is a positive development, since regional authorities are familiar with the needs of

[51] The CBD has designated such areas as subject to Other Effective area-based Conservation Measures (OECMs). This designation was formulated at the tenth international meeting convened by the CBD in 2010.

the individuals, businesses, and municipalities in their respective areas. At the same time, due to a lack of material or human resources within the RCMs, environmental issues, including the protection of natural environments are typically relegated to the back burner. A commission to protect natural land could support RCMs in their land protection work by helping them identify which priority natural environments under their jurisdiction should be protected and integrated into regional protected area networks.

Of course, these solutions would need to be accompanied by a review of the responsibilities of the Quebec Ministry of the Environment, Fight against Climate Change, Wildlife and Parks (MELCCFP). If Quebec is serious about protecting 30% of its land in the interest of biodiversity by 2030, it must increase this strategic ministry's authority and financial resources.

Here are just a few suggestions for what could be achieved in the short term: fund the creation of a commission to protect natural land; implement new financial and environmentally-oriented taxation tools to help municipalities participate in conserving natural environments;[52] increase the tax credits or financial compensation for groups or individuals who donate their land or agree to cede land-use rights in favour of protected areas; and improve the state of knowledge of Quebec's biodiversity.

✦

With greater participation by municipal and regional elected officials, and a realistic strategy based on landscape type, we could very probably achieve the 30% protected area target, and maintain the ecological representativeness and biodiversity of Quebec's land by carefully selecting

[52] The current municipal tax system is an obstacle to the acquisition of land for conservation purposes, since building and land taxes are often municipalities' sole source of funding.

the areas to be conserved according to the two criteria of distinctiveness and vulnerability.

We could also envisage the creation of a network to monitor the ecological integrity of lands by means of eco-acoustics. This would involve setting up a system of standalone listening stations in various parts of Quebec to measure the health of natural environments by compiling data at regular intervals at specific times of the year.

As this book shows, the study of soundscapes provides solid evidence concerning the state of our natural environments. The preservation of pristine soundscapes contributes to the preservation of the whole: a site rich in a multitude of natural soundscapes has greater ecological integrity than one that is silent or polluted by man-made noise.

While ecoacoustics is a complementary tool in the arsenal of biologists whose primary concern is preserving habitats and biodiversity, it is very much underutilized. That's why an extensive acoustic monitoring network in Quebec would greatly complement current scientific knowledge and guide the decisions that need to be made to maintain the province's ecological integrity at an acceptable level.

By Way of Conclusion

In the early 1960s, American biologist Rachel Carson captured worldwide attention with warnings about the harmful effects of DDT,[53] a well-known chlorine-based insecticide. Her concern was that since the substance accumulated and concentrated as it moved up the food chain, the effect was, as she wrote in her book *Silent Spring* (1962), "not even the return of the birds may be taken for granted."

DDT was banned several decades ago and replaced by more targeted substances with fewer negative effects on the health of humans and certain animal populations. However, these other substances have also proved to be extremely lethal in eradicating insects and other arthropods in field crops and other plant habitats.

Habitat loss and fragmentation (together with the other threats mentioned and the degradation caused by widespread pesticide use) are emptying the skies and the Earth of many species that are invisible to our eyes, yet which are essential to proper ecosystem functioning.

Without insects pollinating flowers, there would be no fruit, and without bacteria, fungi, other microorganisms, and subterranean invertebrates tilling and enriching the soil, there would be fewer predators and fewer prey. No vegetation growth means no herbivores, and no herbivores means no carnivores.

[53] DDT: the acronym for dichlorodiphenyltrichloroethan

Apart from the birds not returning, what else can we expect? A continuous series of silent springs?

✦

What principles justify our claim that the presence of *Homo sapiens* on this Earth is of greater value than that of an oak, a bumblebee, or a scallop larva?

With due humility, we need to acknowledge the vanity of the human adventure, given that trees and plants play a far more important ecological role in terrestrial ecosystems than vertebrates (reptiles, amphibians, birds, and mammals, including humans). After all, 99.7% of the planet's biomass is plant-based, and only 0.3% is animal or fungal (mushroom biomass).

What's more, insects and other invertebrates make up the bulk of animal biomass. These little creatures at the primary trophic levels of the food chain, regulate the health of our living environments and are worth their weight in gold. For example, the total weight of all the ants in the world is 20% of that of all humans.

By what standards can we claim that conversations between humans in a crowded restaurant on a Saturday night are worth more than the songs of grey treefrogs at dusk or whales on the high seas?

A little humility is certainly in order then. Let's reduce our impact on the biosphere, and, along the way, take the time to appreciate the beauty offered by Mother Nature, including the symphony of living things, the true music of the world.

In 2020, an extensive survey in the USA revealed that the typical American spends 87% of their time indoors (at home or at work) and 6% in their car. If outings in urban settings, including parks, are excluded, how many hours a year of the remaining 7% of their time does the average American spend in natural environments like forests or lakesides? Shouldn't we be counting in minutes?

One of the most economical and effective public health measures for city dwellers is undoubtedly maintaining the integrity of local natural environments and high-quality natural soundscapes. Plus, getting in touch with nature lets us appreciate its full worth.

✦

As human pressures on the biosphere increase, the environmental question is increasingly central to our lives. That is why crucial and difficult decisions need to be taken now: there is no Plan B, no Planet B. For our own sake and to guarantee the future for those that come after us on this good old Earth, we need to protect more natural habitats.

Like NASA's rocket launches at Cape Canaveral, we have a window of opportunity to undertake this recovery operation. But unlike NASA's launches, which can always be altered when the weather is unfavourable, we have little manoeuvring room when it comes to protecting biodiversity. According to many specialists, if we keep postponing deadlines, the biosphere could reach the point of no return, where the stability of natural systems can no longer be taken for granted. The time to act is now or never.

Of course, when I say that, I sincerely hope I'm wrong.

✦

Like other primates, humans make extensive use of their sight. But as soon as we enter the forest, we lose our customary visual cues; we can't see far, and we can't estimate distances.

It's in the woods that the ear supplants the eye. Listening once again becomes essential—to hear danger coming, to overcome the fear of what we can't see and know, and to keep moving forward, nonetheless.

Is that crack of a branch the sound of a bear coming through the trees at the edge of the woods? Is the constant rustling of poplar leaves over there on the hilltop an indication of a malevolent predator?

Natural patterns of communication via sound are being transformed by climate change, our ever-expanding road networks, and the loss and degradation of wildlife habitats. The balance of factors governing ecosystem biodiversity is extremely delicate and fragile.

It is remarkable how the voices of various species in the acoustic spectrograms of healthy environments come out so distinctly, each with its own place, pitch, period, and tempo.

When the ecosystem breaks down, the natural harmony of nature's orchestra is lost, disorder sets in, and the balance and diversity of voices disappear. The relationship between the intensity of a species' signal and background noise is no longer the same. The narrowness of natural habitats and the growing din of human societies are converging to overwhelm the acoustic communication needed for wild species' maintenance and reproduction.

✦

Imagine a sound mixing console at the back of a symphony hall. Imagine that you're the hall's sound engineer, and that at one time you had twenty-four soundtracks at your disposal, reproducing with subtlety and nuance at least twenty-four instrument groups—cellos, clarinets, bassoons, xylophones, etc.—or individual instruments like triangles, timpanis, or first violins. Unfortunately, due to recent budget cuts, you're down to only six or eight tracks. You can no longer reproduce subtle timbres and nuances. As if that weren't enough, the audience is getting louder and louder, with no regard for the conductor or the struggling musicians on stage. This naturally affects the depth and quality of the orchestra's performances.

We're in quite a different world these days, living in an environment that no longer sounds like it did 100, 200, or 1,000 years ago. A world where we're gradually simplifying the musical score and amplifying the background noise.

✦

Millennia ago, our hunter-gatherer ancestors had to have accurate knowledge of their soundscapes: their survival depended on it. Have we, their descendants, lost our awareness of the sounds of rumbling thunder, of prey to be captured, and of growling predators? Is this knowledge still buried somewhere within us?

Each of us experiences the world's soundscapes through a personal filter. Our senses, in general, are not simply windows through which we perceive our environment, they also serve as magnifying or distorting lenses that enable us to hear, see, smell, taste, and touch according to the personal and cultural baggage of our beliefs, mindsets, age, gender, and places of birth. All human beings generally have the same capacities to listen and speak, yet we use hundreds of different languages and dialects. The cultural use of listening varies widely insofar as we "interpret" the environment according to distinct cultural and personal repertoires.

Overall then, our soundscapes are all the more multiple, diverse, and magnificent when the voices of our own species down through time are added to the concert of the other living species with which we share the planet, especially when we respect the original score.

✦

Our species' love of music may be a key or trigger to an awareness of our disproportionate hold over nature's orchestra. Our species makes too much noise and like a timpani player who plays too loud, doesn't follow their assigned score, and refuses to stop when the conductor orders them to, with the result that the music we perform lacks subtlety.

Attraction to music is one of humanity's most universal traits.

Music in a variety of forms is omnipresent in all cultures and among all peoples. In every period of our history, it has accompanied joyous events from the celebration of a successful hunt or a monarch's wedding to the marking of tragic occasions, such as a loved one's funeral.

But why do we love music in the first place? For evolutionary specialists, this remains a riddle: why does our species devote so much time and energy to this activity which, at first blush, seems to have no real purpose? If music is the result of our species' adaptation through natural selection, what could possibly be the benefit for our ancestors to have perpetuated it from generation to generation? For while our everyday gestures and activities, such as eating, speaking, and making love, have a biological purpose and have been shaped by natural selection in order to survive, exchange information, and reproduce so that our heritage can be passed on to the next generation, playing and listening to music does not, strictly speaking, seem to have a purpose.

In the late 1970s, biologists Stephen J. Gould and Richard C. Lewontin hypothesized that certain evolutionarily useless traits, like our interest in and aptitude for music, might simply be the collateral effect of other traits of more practical benefit for species survival and perpetuation.

Music, for example, has much in common with language. Certain elements in the mastery of language are closely associated with similar parameters in the appreciation and comprehension of musical language. The right hemisphere of the human brain processes and recognizes pitch-related information, while the left hemisphere processes and recognizes rhythm-related information. Human children develop their language and musical skills very early in life. From the age of five months, they can recognize melodies even when transposed into a different key, i.e., when played from a different starting note and on a different scale, but respecting the intervals. They can also recognize a melody even if its tempo is faster or slower.

Our musical abilities may also be linked to the need to differentiate between ambient sounds that indicate danger and those that are neutral or even positive, such as those indicating prey nearby. Neurologists and psychologists have suggested that fear-inducing music may be a throwback to an ancestral defence mechanism against a threat, whereas a pleasant piece of music may be associated with memories of sounds that generate feelings of happiness, such as opportunities to socialize. The content of a sound stimulus, whether positive or negative, influences the secretion of well-being-associated hormones like endorphins or stress-associated hormones like testosterone.

Sound is a powerful means of reconnecting with the natural world, because it triggers emotions. It inserts itself physically within us. Its waves caress or hit against our skin, strike the structures of our inner ear, and pass through our internal organs and bones. It revives deeply buried memories and reminds us of happy (and sometimes unhappy) times shared with loved ones in cherished settings.

Music has a social function: it facilitates and supports the sharing of these emotions with others. It can unite individuals and reduce interpersonal tension. Viewed through the filter of evolution, music can be considered a useful adaptation because it can improve communication, cohesion, and solidarity among members of a given group. In short, it trains humans from an early age to be sociable.

For philosophers Friedrich Nietzsche and Arthur Schopenhauer, music was nothing less than the voice of the cosmos.

✦

R. Murray Schafer, the theorist mentioned at the start of this book, wrote that humans like to make noise to remind themselves that they are not alone in the universe, and to remind others that they exist.

Couldn't we try to be more attentive to the sounds made by the other living beings with whom we share the planet? Does this human love of music open a door that could enable us to better protect natural environments? Could awareness of the natural world's beautiful soundscapes lead to greater respect for nature?

As mentioned in Chapter 5, the ear is not only the organ of hearing, but also that of balance. More broadly, our place in the biosphere is also a question of balance.

✦

Imagine a string quartet: cello, viola, violin, and double bass. Then imagine that each instrument represents a different taxonomic group within the living world: the double bass by fungi, bacteria, viruses, and the other humble single-cell organisms at the base of ecosystems; the cello by plants; the viola by all other non-human animals, and the violin by humans, given their customary penchant for the most flamboyant score.

Shouldn't we reverse the roles and redistribute the instruments according to the actual role of the instrument players in the natural environment? Mushrooms and other fungi should play the violin, while we, humans, would have to make do with the double bass. While we're at it, why not replace the double bass with a simple triangle, given our insignificant role in nature's mechanics?

According to Schafer, the world is an immense musical composition without beginning or end. To most intents and purposes, we are currently the composers of that music—and we can make it better or worse.

✦

Let's not forget that we live in a green world: 99.7% of the planet's biomass is plant-based.

Trees are the most enduring organisms on the planet: their life expectancy far exceeds not only our own but also that of all other animal species. Trees go back to the extremely distant past (the first conifers appeared 300 million years ago, and deciduous trees 135 million years ago). They are also extremely diverse (there are now an estimated 73,000 tree species, including 9,200 still unknown to science). Thanks to natural selection, trees have also developed many mechanisms to receive, process, and exchange sensory information about their environment.

Will we ever know what trees hear? And if they can hear, can they reproduce sounds? And if they can reproduce sounds, will we ever know whether they can sing?

EPILOGUE

As I sit here near the Ouareau River rapids in the Lanaudière region north of Montreal, snatches of discussions during COP15 in December 2022 come to mind. Isn't it remarkable that almost two hundred countries on that occasion pledged to protect nature and maintain sufficient natural habitats to ensure the world's future? Or was all that just more wishful thinking? An occasion for high-level UN diplomatic pirouettes with no real intention of solving the problem?

One thing is certain: the efforts required to turn words into action are colossal. Unfortunately, a comparison in terms of boldness and scope between the Kunming-Montreal Global Biodiversity Framework at COP15 and the Paris Climate Agreement of 2015 raises doubts. While the latter was just as ambitious with its overriding goal of limiting global warming to 1.5°C above pre-industrial levels, we must admit that the commitment made by almost all the nations of the world in Paris is still, for most of them, just words on a piece of paper.

The river waters undulate, froth, and noisily flow by. At 69.5 dB, the sound is as loud as traffic on a city boulevard. Yet the sound of the river soothes me because it is natural and surprisingly constant.

Ever since the last glacier left the shores of Upper America millennia ago, leaving behind enormous rocks that shape the rapids in this section of the river, the river itself flows unperturbedly south towards the St. Lawrence River. Season after season, year after year, even in the depths of winter, the river flows down the same slope at more or less the same pace, because the current is strong enough to prevent the ice from diminishing the river's ardour.

As I wrote at the beginning of this book, all too often we only become interested in something when we're about to lose it. I also confided that my hearing has become worse with age. That's a big source of worry for me: I dread the day when I'll no longer be able to savour the joy of hearing the morning concert of birds in the deep forest.

It's funny how things turn out because it was my father who instilled in me an interest in birds, plants, insects, and everything else that lives, breathes, and moves in nature. A sailor and mechanic by trade, my father damaged his ears in the holds of the ships on which he worked. When he was around the same age as I am now, he lost much of his hearing. He doubtless loved birds as much as I did.

I distinctly remember the time he took me to the edge of a nearby forest just across from Île d'Orléans on the south shore of the St. Lawrence River. I was eight or nine years old at the time, and it was as a family camping trip was coming to an end.

After placing a light wooden cage in the undergrowth, he told me to hide with him in a thicket, signalling with an index finger across his mouth that we had to be absolutely silent.

My father liked to tinker with little artificial cages. They had a tipping mechanism in the form of a board on a pivot; he used to place tasty seeds rich in fat and protein on the board. When a bird landed on the board, its weight, no matter how light, triggered the cage door to close automatically. However, my father's purpose wasn't to capture birds; he was simply interested in perfecting his gizmo and observing which species of bird were most likely to take the bait.

On this particular occasion, it was a small, brownish bird that was seduced by the morsels on offer, and after circling the contraption for a long time, finally landed on the board, and as expected, the cage door slammed shut behind it.

We both stood up as one.

"Come on," said my father, "let's see what we've caught."

It looked like a brown-striped sparrow. It could've been mistaken for a house sparrow, were it not for its white throat, elegant yellow eyebrow, and small, black-and-white-striped head. In one swift movement, my father opened the cage and caught the bird in his right hand, then, with infinite gentleness, placed it in the palm of my hands.

"It's a Peabody," he added, "the bird that goes: *Old Sam, Peabody, Peabody, Peabody!!*

I later learned the bird's common name, white-throated sparrow, and, even later, its scientific name, *Zonotrichia albicollis*. I could feel its heartbeat and the radiant warmth of its body in the palms of my hands.

I looked at it for a long, long time.

"Should we let him go?" my father asked.

I opened my hands and the bird immediately flew away.

We didn't hear the white-throated sparrow singing that day. But remembering my father's description, I have always recognized that familiar song from the edge of the woods.

As I mentioned, the famous *Old Sam, Peabody, Peabody, Peabody!* is heard less often now, as males who shorten the final triplet to *Old Sam, Peabo, Peabo, Peabo!* are apparently more popular with females. It's believed that the new version has gone viral on the white-throated sparrows' wintering grounds, with the vast majority succumbing to the lure of novelty.

In my case, it's this bird's traditional call that is imprinted on my memory. Forever. Because with age, we become more impervious to passing fads.

Few of my childhood memories are as clear as this one.

While I may lose my hearing in the medium or long term, a few of the sounds that have marked my life will stay with me until my last breath, including the roar of the Ouareau rapids and the song of the white-throated sparrow, *Old Sam, Peabody, Peabody, Peabody!*

Acknowledgements

Many thanks to all my initial reviewers: Sandra Sultana, for taking on the unenviable task of commenting on the manuscript's first draft and pointing out its imperfections; Pierre Bonneau, for his attentive and thorough reading of the text; Benoît Archambault, singer-songwriter and storyteller, for his creative suggestions on how to enliven the text and extend its life via other forms; Thibaut Quinchon, sound designer and acoustic ecology enthusiast, for our long discussions on the book's themes, as well as for all our joint projects, both current and future, and, above all, for his infectious energy; and Raphaël Proulx, research professor in the Biological Sciences Department at the Université du Québec à Trois-Rivières, for his feedback and especially for providing such a relevant preface.

Many thanks also to the entire Éditions MultiMondes team, led by Dominique Lemay and Raymond Lemieux, for their advice, support, and meticulous work at every stage of production from layout to marketing.

I, of course, remain solely responsible for the ideas presented in the text and for any errors that may have crept in despite my careful revision of content.

Last but not least, many thanks to all those wonderful people who work, mostly unseen and often as volunteers, in small or large organizations dedicated to studying and protecting wild species, and conserving the Earth's natural environments in as pristine a state as possible.

Bibliography

Andrén, H., "Effects of habitat fragmentation on birds and mammals in landscapes with different proportions of suitable habitat: a review," *Oïkos* no. 71, 1994, 355-366.

Appel, H.M., and R.B. Cocroft, "Plants respond to leaf vibrations caused by insect herbivore chewing," *Oecologia* no. 175, 2014, 1257-1266, https://doi:10.1007/s00442-014-2995-6.

Arsenault, D., L. Gagnon, C.A. Martijn, and A. Watchman, "Le projet Nisula : recherche pluridisciplinaire autour d'un site à pictogrammes (DeEh-1) en Haute-Côte-Nord," *Paléo-Québec* no. 23, 1995, 17-57.

Attali, J., *Bruits.* P.U.F., 1977.

Auzel, P., B. Caillié, J. Dupras, et al., *Un Plan Sud pour le Québec : livre blanc pour la protection de la biodiversité au sud du 49e parallèle,* Centre de la science de la biodiversité du Québec, 2021.

Bacon, J., *Bâtons à message,* Éditions Mémoire d'encrier, 2009.

Barnosky, A.D., E.A. Hadley, J. Bascompte, et al., "Approaching a state shift in Earth's biosphere ," *Nature* no. 486, 2012, 52-58, https://doi:10.1038/nature11018.

Bastien, R., T. Bohr, B. Moulia, and S. Douady, "Unifying model of shoot gravitropism reveals proprioception as a central feature of posture control in plants," *Proceedings of the National Academy of Sciences of the United States of America* no. 110 (2), 2012, 755-760, https://doi.org/10.1073/pnas.1214301109.

Bengtson, S.A., "Breeding ecology and extinction of the great auk (Pinguinus impennis): anecdotal evidence and conjectures," *The Auk* no. 101 (1), 1984, 1-12.

Birdlife International, *Threatened Birds of the World*, Lynx Nature books and Birdlife International, 2000.

Bibeau, G., *Les Autochtones – la part effacée du Québec,* Éditions Mémoire d'encrier, 2020.

Blickley, J.L., K.R. Word, A.H. Krakauer, et al., "Experimental chronic noise is related to elevated fecal corticosteroid metabolites in lekking Male Greater Sage-Grouse (*Centrocercus urophasianus*)," *PLoS ONE* no. 7 (11), 2012, e50462, https://doi:10.1371/journal.

pone.0050462.

Blumberg, M.S., G. Sokoloff, R.F. Kirby, and K.J. Kent, "Distress vocalizations in infant rats: what's all the fuss about ?," *Psychological Science* no. 11 (1), 2000, 78-81.

Boyd, R., and J. Silk, *L'aventure humaine, des molécules à la culture,* De Boeck et Larcier, 2004.

Bunkley, J.P., C.J.W. McClure, A.Y. Kawahara, C.D. Francis, and J.R. Barber, "Man-made noise changes arthropod abundances," *Ecology and Evolution* no. 7, 2017, 2977-2985, https://doi.org/10.1002/ece3.2698.

Buxton, R.T., A.L. Pearson, C. Allou, K. Fristrup, and G. Wittemyer, "A synthesis of health benefits of natural sounds and their distribution in national parks," *Proceedings of the National Academy of Sciences of the United States of America* no. 118 (14), 2021. https://doi.org/10.1073/pnas.2013097118.

Carson, R., *Silent Spring*, Houghton Mifflin Company, 1962.

Clark, A. and D.J. Chalmers. "The Extended Mind," *Analysis* no. 58, 1988, 7-19.

COSEPAC, "Évaluation et Rapport de situation du COSEPAC sur le courlis esquimau (*Numenius borealis*) au Canada," Comité sur la situation des espèces en péril au Canada, 2009, vii + 35

COSEPAC, "Évaluation et Rapport de situation du COSEPAC sur la rainette faux-grillon de l'ouest (*Pseudacris triseriata*) population carolinienne et population des Grands Lacs et Saint-Laurent et du Bouclier canadien au Canada – Mise à jour," Comité sur la situation des espèces en péril au Canada, 2008, vii + 55

Cowie, R.H., P. Bouchet, and B. Fontaine, "The sixth mass extinction: fact, fiction or speculation," *Biological Reviews* no. 97, 2022, 640-663, https://doi.org/10.1111/brv.12816.

Creel, S., J.E. Fox, A. Hardy, J. Sands, B. Garrott, and R.O. Peterson, "Snowmobile activity and glucocorticoid stress responses in wolves and elk," *Conservation Biology* no. 16 (3), 2002, 809-814.

David, B., *À l'aube de la 6e extinction : comment habiter la Terre*, Éditions Grasset and Fasquelle, 2021.

Faure, A., and R.R. Hoy, "The sounds of silence: cessation of singing and song pausing are ultrasound-induced acoustic startle behaviors in the katydid *Neoconocephalus ensiger* (Orthoptera; Tettigoniidae)," Journal of *Comparative Physiology* no. 186, 2000, 129-142.

Gagliano, M., "Green symphonies: a call for studies on acoustic communication in plants," *Behavioral Ecology* no. 24 (4), 2013, 789-796, https://doi:10.1093/beheco/ars206.

Gagliano, M., M. Grimonprez, M. Depczynski, and M. Renton, "Tuned in: plant roots use sound to locate water," *Oecologia* no. 184, 2017, 151–160, https://doi.org/10.1007/s00442-017-3862-z.

Gagliano, M., M. Renton, M. Depczynski, and S. Mancuso, "Experience teaches plants to learn faster and forget slower in environments where it matters," *Oecologia* no. 175,

2014, 63-72. https://doi.org/10.1007/s00442-013-2873-7.

Gagliano, M., S. Mancuso, and D. Robert, "Towards understanding plant bioacoustics," *Trends in Plant Science* no. 17 (6), 2012, 323-325, https://doi.org:10.1016/j.tplants.2012.03.002.

Gatti, R.C., P.B Reich, J.G.P. Gamarra, and J. Liang, "The number of tree species on Earth," *Proceedings of the National Academy of Sciences of the United States of America no.* 119 (6), 2022, https://doi.org/10.1073/pnas.2115329119.

Gosselin, M., and M. Robert, Tourte. In J. Gauthier, and Y. Aubry (eds.), *Les oiseaux nicheurs du Québec : Atlas des oiseaux nicheurs du Québec méridional*, Association québécoise des groupes d'ornithologues, Société québécoise de protection des oiseaux and Canadian Wildlife Service, Environment Canada, Quebec Region, 1995, 1158-1159.

Gould, S. J., and R.C. Lewontin, "The spandrels of San Marco and the Panglossian paradigm: a critique of the adaptationist programme," *Proceedings of the Royal Society of London, Series B*, no. 205 (1161), 1979, 581-598.

Hallé, F., *Plaidoyer pour l'arbre*, Actes Sud, 2005.

Hebert, Y., "John James Audubon au Québec," *Nature sauvage* no. 36 (3), 2017, 16-21.

Jones, M.T., and L.L. Willey, *Biology and Conservation of the Wood Turtle,* Northeast Association of Fish and Wildlife Agencies, 2021.

Jung, M., A. Arnell, X. De Lamo, et al, "Areas of global importance for conserving terrestrial biodiversity, carbon and water," *Nature Ecology and Evolution* no. 5, 2021, 1499-1509, https://doi.org/10.1038/s41559-021-01528-7.

Kanapé Fontaine, N., *Manifeste Assi,* Éditions Mémoire d'encrier, 2014.

Karban, R., *Plant Sensing and Communication*, The University of Chicago Press, 2015.

Keenan, T.F., A.D. Richardson, and K. Hufkens, "On quantifying the apparent temperature sensitivity of plant phenology," *New Phytologist* no. 225 (2), 2020, 1033-1040, https://doi.org/10.1111/nph.16114.

Krause, B., *Le grand orchestre animal*, Flammarion, 2013.

Krause, B., "The Niche Hypothesis: A virtual symphony of animal sounds, the origins of musical expression and the health of habitats," *The Explorers Journal*, 1993, 156-160.

Leakey, R., and R. Lewin, *La Sixième extinction : évolution et catastrophes,* Flammarion, 1997.

Leboeuf, M., C. Chouinard, R. Bourgeois, and D. Grenier-Héon, "Effets des infrastructures linéaires sur la biomasse des insectes nocturnes à l'échelle du paysage," *Le Naturaliste canadien* no. 145 (2), 2021, 68-78.

Leboeuf, M., *Le dernier caribou*, Éditions MultiMondes, 2020.

Leboeuf, M., *Paroles d'un bouleau jaune*, Éditions MultiMondes, 2018.

Leboeuf, M., *Arbres et plantes forestières du Québec et des Maritimes*, 2nd ed., Éditions Michel Quintin, 2016.

Leboeuf, M., *Nous n'irons plus au bois : la biodiversité québécoise en péril*, Collection Nature sauvage, Vélo Québec éditions, 2010.

Legris, M., B.M. Szarzynska-Erden, M. Trevisan, L. Allenbach-Petrolati, and C. Fankhauser, "Phototropin-mediated perception of light direction in leaves regulates blade flattening," *Plant Physiology* no. 187, 2021, 1235-1249, https://doi.org/10.1093/plphys/kiab410.

Leopold, A., *Almanach d'un comté des sables,* Flammarion, 2000.

Linteau, P.A., R. Durocher, J.C. Robert, and F. Ricard, *Histoire du Québec contemporain,* tome II – Le Québec depuis 1930, Éditions du Boréal, 1989.

Locke, H., E.C. Ellis, O. Venter, et al., "Three global conditions for biodiversity conservation and sustainable use: an implementation framework," National Science Review no. 6 (6), 2019, https://doi.org/10.1093/nsr/nwz136.

MacArthur, R.H., "Population ecology of some warblers of northeastern coniferous forests," *Ecology* no. 39 (4), 1958, 599-619.

Margulis, L., and D. Sagan, *What is Life?*, California University Press, 1995.

Maeder, M., M.M. Gossner, A. Keller, and M. Neukom, "Sounding soil: an acoustic, ecological and artistic investigation of soil life," *Soundscape* no. 18, 2019, 5-14.

Marie-Victorin, Frère, *Croquis laurentiens*, Éditions Fides, 1982.

Matsuhashi, M., et al., "Production of sound waves by bacterial cells and the response of bacterial cells to sound," *Journal of General and Applied Microbiology*, no. 44 (1), 1998, 49-55, https://doi.org/10.2323/jgam.44.49.

McKenna, M. F., G. Shannon, and K. M. Fristrup, "Characterizing man-made noise to improve understanding and management of impacts to wildlife," *Endangered Species Research* no. 31, 2016, 279-291.

Mikkola, H., *Owls of the World*, 2nd ed., Firefly Books, 2014.

Mousette, M., and G.A. Waselkov, *Archéologie de l'Amérique coloniale française*, Lévesque Éditeur, 2013.

Muñoz, P.T., F.P. Torres, and A.G. Megias, "Effects of roads on insects: a review," *Biodiversity and Conservation* no. 24, 2015, 659-682.

Nakano, R. T. Takanachi, T. Fujii, N. Skals, A. Surlykke, and Y. Ishikawa, "Moths are not silent, but whisper ultrasonic courtship songs," *Journal of Experimental Biology* no. 212 (24), 2009, 4072-4078.

Naughton, D., *Histoire naturelle des mammifères du Canada*, Éditions Michel Quintin and Canadian Museum of Nature, 2016.

Niinemets, U., and M. Reichstein, "Controls on the emission of plant volatiles through stomata: a sensitivity analysis," *Journal of Geophysical Research* no.108, 2003, https://doi.org/10.1029/2002JD002626.

O'Grady, C., "Warming of 1.5°C carries risk of crossing climate tipping points," *Science* no. 377 (6611), 2022, https://doi/.org/10.1126/science.ade7662.

Ohno, M, and A. Yamawo, "Night interruption provides evidence for photoperiodic regulation of bud burst in Japanese beech," *Fagus crenata, Plant Signaling & Behavior* no. 16 (12), 2021, https://doi.org/10.1080/1559232
4.2021.1982562.

Otter, K.A., A. McKenna, S.E. Lazerte, and S.M. Ramsay, "Continent-wide shifts in song dialects of White-throated Sparrows," *Current Biology* no. 30 (19), 2020, https://doi.org/10.1016/j.cub.2020.05.084.

Paillé, G., *Histoire forestière du Québec*, Les Publications du Québec, 2012.

Parise, A.G., M. Gagliano, and G.M. Souza, "Extended cognition in plants: is it possible?," *Plant Signaling & Behavior* no. 15 (2), 2020, e1710661, https://doi.org/10.1080/15592324.2019.1710661.

Pérez-González, J., G. Rey Gozalo, D. Montez González, S.J. Hidalgo de Trucios, and J.M. Barrigón Morillas, "Are quartzite scree slopes used by birds to promote sound transmission in the Mediterranean forest?," *Animal Biodiversity and Conservation* no. 44 (2), 2021, https://doi.org/10.32800/abc.2021.44.0175.

Pintal, J.Y., J. Provencher, and G. Piedalue, *Archéologie du Québec : territoire et peuplement*, Musée Pointe-à-Callière and Les Éditions de l'Homme, 2015.

Phillips, J.N., S.E. Termondt, and C.D. Francis, "Long-term noise pollution affects seedling recruitment and community composition, with negative effects persisting after removal," *Proceedings of the Royal Society* B no. 288, 2021, https://doi.org/10.1098/rspb.2020.2906.

Recio, A., C. Linares, J.R. Banegas, and J. Diaz "The short-term association of road traffic noise with cardiovascular, respiratory, and diabetes-related mortality," *Environmental Research* no. 150, 2016, 383-390.

Reece, J., L. Urry, M. Cain, S. Wasserman, P. Minorsky, and R. Jackson, *Campbell Biology*, 9th ed., Pearson Benjamin Cummings, 2011.

Reijnen, R., and R. Foppen, "The effects of car traffic on breeding bird populations in woodland: evidence of reduced habitat quality of willow warblers (*Phylloscopus trochilus*) breeding close to a highway," *Journal of Applied Ecology* no. 31, 1994, 85-94.

Rioux, J.F., *Évolution de la couverture terrestre du Québec méridional sous l'influence des grands centres urbains.* Government of Quebec, Institut de la statistique du Québec, 2019, 23.

Robert, M., and M. Gosselin, Canard du Labrador. In J. Gauthier, and Y. Aubry (eds.), *Les oiseaux nicheurs du Québec : Atlas des oiseaux nicheurs du Québec méridional*, Association québécoise des groupes d'ornithologues, Société québécoise de protection des oiseaux and Canadian Wildlife Service, Environment Canada, Quebec Region, 1995, 1111.

Robert, M., and M. Gosselin, Grand pingouin. In J. Gauthier, and Y. Aubry (eds.), *Les oiseaux nicheurs du Québec : Atlas des oiseaux nicheurs du Québec méridional,* Association québécoise des groupes d'ornithologues, Société québécoise de protection des oiseaux and

Canadian Wildlife Service, Environment Canada, Quebec Region, 1995, 1156.

Saucier, J.-P. et al, "Écologie forestière." In *Ordre des Ingénieurs forestiers du Québec, Manuel de foresterie*, 2nd ed., Éditions MultiMondes, 2009, 165-315.

Schafer, R.M., *The Tuning of the World: The Soundscape*, Alfred A. Knopf, 1977.

Senzaki, M., Y. Yamaura, C.D. Francis, and F. Nakamura, "Traffic noise reduces foraging efficiency in wild owls," *Scientific Reports* no. 6, 2016, https://www.doi.org/10.1038/srep30602.

Sibley, D.A., Birding Basics, Alfred A. Knopf, 2002.

Sorensen, M., M. Hvi Dberg, Z.J. Andersen, et al., "Road traffic noise and stroke: a prospective cohort study," *European Heart Journal* no. 32 (6), 2011, 737-744.

Sueur, J., *Histoire naturelle du silence*, Actes Sud, 2023.

Thiltges, S. "Une écologie du chant d'oiseaux : poétique de l'écoute dans l'Arrière-saison d'Adalbert Stifter." In *La forêt sonore : de l'esthétique à l'écologie.* J. Mottet and F. Deroussen (eds.), Éditions Champ Vallon, 2017, 198-213.

Thoreau, H.D., *Walden ou la vie dans les bois,* Gallimard, 1990 (1922).

Truax, B., *Handbook for Acoustic Ecology,* World Soundscape Project (Simon Fraser) and ARC Publications, 1999 (1978),
http://www.sfu.ca/sonic-studio-webdav/handbook/index.html.

Veits, M., I. Khait, U. Obolski, et al., "Flowers respond to pollinator sound within minutes by increasing nectar sugar concentration," *Ecology Letters* no. 22 (9), 2019, 1483-92, https://doi.org/10.1111/ele.13331.

Vélez, A., N.M. Gordon, and M.A. Bee, "The signal in noise: acoustic information for soundscape orientation in two North American treefrogs," *Behavioral Ecology* no. 28 (3), 2017, https://doi.org:10.1093/beheco/arx044.

Villard, M.-A., M. Mazerolle, and S. Haché, "L'impact des routes, au-delà des collisions : le cas des oiseaux forestiers et des amphibiens," *Le Naturaliste canadien* no. 136 (2), 2012, 61-65.

Zohner, C.M., and S.S. Renner, "Perception of photoperiod in individual buds of mature trees regulates leaf-out," *New Phytologist* no. 208, 2015, 1023-30, https://doi.org/10.1111/nph.13510.

Further Listening

Quebec soundscapes:

https://www.uneviequisecoute.com/

Atlas of the St. Lawrence River's marine soundscape:

https://soundscape-atlas.uqar.ca/

Soundscapes from elsewhere in the world:

https://www.wildsanctuary.com/

https://centerforglobalsoundscapes.org/

https://earth.fm/ https://www.soundcities.com/index.php#

Animal sound banks:

https://www.macaulaylibrary.org/ https://sonotheque.mnhn.fr/

Research group websites:

https://ear.cnrs.fr/

https://sites.google.com/site/ecoacousticssociety/

https://research.ecosounds.org/ https://neuropsi.cnrs.fr/departements/cnn/equipe-isabelle-charrier/

https://www.geobotanik.uni-freiburg.de/MasterLab

https://cmast.ncsu.edu/people-at-cmast/eggleston/outreach/soundscapes/

Correlations Between Common Names and Scientific Names

A

American beaver	*Castor canadensis*
American beech	*Fagus grandifolia*
American black bear	*Ursus americanus*
American burying beetle	*Nicrophorus americanus*
American crow	*Corvus brachyrhynchos*
American dipper	*Cinclus mexicanus*
American hornbeam	*Carpinus caroliniana*
American marten	*Martes americana*
American mountain ash	*Sorbus americana*
American robin	*Turdus migratorius*
American three-toed woodpecker	*Picoides dorsalis*
American woodcock	*Scolopax minor*
Arctic char	*Salvelinus alpinus*
Atlantic puffin	*Fratercula arctica*

B

Balsam fir	*Abies balsamea*
Balsam poplar	*Populus balsamifera*
Barred owl	*Strix varia*
Basswood	*Tilia americana*
Bay-breasted warbler	*Setophaga castanea*
Beluga	*Delphinapterus leucas*

Bitternut hickory	*Carya cordiformis*
Black ash	*Fraxinus nigra*
Black maple	*Acer nigra*
Black spruce	*Picea mariana*
Black-backed woodpecker	*Picoides arcticus*
Blackburnian warbler	*Setophaga fusca*
Black-capped chickadee	*Poecile atricapillus*
Black-chokeberry	*Aronia melanocarpa*
Black-tailed prairie dog	*Cynomys ludovicianus*
Black-throated blue warbler	*Setophaga caerulescens*
Bloodroot	*Sanguinaria canadensis*
Blue jay	*Cyanocitta cristata*
Blue-headed vireo	*Vireo solitarius*
Bobcat	*Lynx rufus*
Bog Labrador tea	*Ledum groenlandicum*
Boreal chickadee	*Poecile hudsonicus*
Boreal chorus frog	*Pseudacris maculata*
Broad-winged hawk	*Buteo platypterus*
Brook trout	*Salvelinus fontinalis*
Brown bear	*Ursus arctos*
Brown thrasher	*Toxostoma rufum*

C

California scrub jay	*Aphelocoma californica*
Canada goose	*Branta canadensis*
Canada jay	*Perisoreus canadensis*
Canada lynx	*Lynx canadensis*
Canada warbler	*Cardellina canadensis*
Cerulean warbler	*Setophaga cerulea*
Chimney swift	*Chaetura pelagica*
Cinereous shrew	*Sorex cinerus*
Colorado pinyon	*Pinus edulis*

Common blackbird	*Turdus merula*
Common fruit fly	*Drosophila melanogaster*
Common loon	*Gavia immer*
Common raven	*Corvus corax*
Common reed	*Phragmites australis* spp. Australis
Common tree cricket	*Oenothera biennis*
Corn	*Zea mays*
Cougar	*Puma concolor*
Coyote	*Canis latrans*
Cucumber root	*Medeola virginiana*

D

Deer mouse	*Peromyscus maniculatus*
Dog	*Canis*
Dovekie	*Alle alle*

E

Eastern grey squirrel	*Sciurus carolinensis*
Eastern hemlock	*Tsuga canadensis*
Eastern meadowlark	*Sturnella magna*
Eastern wood peewee	*Contopus virens*
Eskimo curlew	*Numenius borealis*
Eurasian blackcap	*Sylvia atricapilla*
Eurasian blue tit	*Cyanistes caeruleus*
Eurasian chaffinch	*Fringilla coelebs*
European buckthorn	*Rhamnus catharticus*
European robin	*Erithacus rubecula*

F

Field sparrow	*Spizella pusilla*

G

Garlic mustard	*Alliaria petiolate*
Glandular birch	*Betula glandulosa*
Golden eagle	*Aquila chrysaetos*
Golden-crowned kinglet	*Regulus satrapa*
Grey treefrog	*Hyla versicolor*
Great auk	*Pinguinus impennis*
Great blue heron	*Ardea herodias*
Great crested flycatcher	*Myiarchus crinitus*
Great grey owl	*Strix nebulosa*
Great tit	*Parus major*
Greater yellowlegs	*Tringa melanoleuca*
Great-horned owl	*Bubo virginianus*
Green lacewings	*Chrysopidae*

H

Hackberry	*Celtis occidentalis*
Hermit thrush	*Catharus guttatus*
Hooded seal	*Cystophora cristata*
Horned lark	*Eremophila alpestris*
House sparrow	*Passer domesticus*
Humpback whale	*Megaptera novaeangliae*

J

Jack pine	*Pinus banksiana*
Japanese beetle	*Popillia japonica*
Japanese knotweed	*Fallopia japonica*

L

Labrador duck	*Camptorhynchus labradorius*
Large-flowered bellwort	*Uvularia grandiflora*
Long-eared owl	*Asio otus*

Long-finned pilot whale	*Globicephala melas*
Long-tailed weasel	*Musleta frenata*

M

Moose	*Alces americanus*
Mourning dove	*Zenaida macroura*

N

Narwhal	*Monodon monoceros*
Nashville warbler	*Leiothlypis ruficapilla*
Northern river otter	*Lontra canadensis*
Northern gannet	*Morus bassanus*
Northern long-eared myotis	*Myotis septentrionalis*
Northern parula	*Setophaga americana*
Northern red oak	*Quercus rubra*
Northern short-tailed shrew	*Blarina brevicauda*
Northern white pine	*Pinus strobus*

O

Orca	*Orcinus orca*

P

Passenger pigeon	*Ectopistes migratorius*
Pea	*Pisum sativum*
Plainfin midshipman	*Porichthys notatus*

R

Razorbill	*Alca torda*
Red maple	*Acer rubrum*
Red spruce	*Picea rubens*
Red squirrel	*Tamiasciurus hudsonicus*
Red-breasted nuthatch	*Sitta canadensis*

Red-eyed vireo	*Vireo olivaceus*
Red-winged blackbird	*Agelaius phoeniceus*
Ring-billed gull	*Larus delawarensis*
Rock dove	*Columba livia*
Rose-breasted grosbeak	*Pheucticus ludovicianus*
Rusty blackbird	*Euphagus carolinus*

S

Scarlet tanager	*Piranga olivacea*
Shagbark hickory	*Carya ovata*
Sharp-lobed hepatica	*Anemone acutiloba*
Sharp-shinned hawk	*Accipiter striatus*
Sharp-tailed grouse	*Tympanuchus phasianellus*
Sheep laurel	*Kalmia angustifolia*
Short-beaked common dolphin	*Delphinus delphis*
Snow goose	*Anser caerulescens*
Snowshoe hare	*Lepus americanus*
Snowy tree cricket	*Oecanthus fultoni*
Song sparrow	*Melospiza melodia*
Southern red-backed vole	*Myodes gapperi*
Soybean	*Glycine max*
Speckled alder	*Alnus incana* ssp. *Rugosa*
Sperm whale	*Physeter macrocephalus*
Spring peeper	*Pseudacris crucifer*
Striped bass	*Morone saxatilis*
Striped maple	*Acer pensylvanicum*
Sugar maple	*Acer saccharum*
Swainson's thrush	*Catharus ustulatus*
Sword-bearing coneheads	*Catharus ustulatus*

T

Tamarack	*Larix laricina*
Tasmanian devil	*Sarcophilus harrisii*
Thale cress	*Arabidopsis thaliana*
Two-leaved mitrewort	*Mitella diphylla*
Two-leafed toothwort	*Cardamine diphylla*

W

Walrus	*Odobenus rosmarus*
White ash	*Fraxinus americana*
White oak	*Quercus alba*
White spruce	*Picea glauca*
White-beaked dolphin	*Lagenorhynchus albirostris*
White-breasted nuthatch	*Sitta carolinensis*
White-footed mouse	*Peromyscus leucopus*
White-throated sparrow	*Zonotrichia albicollis*
Willow warbler	*Phylloscopus trochilus*
Winter wren	*Troglodytes hiemalis*
Wolf	*Canis lupus*
Wolverine	*Gulo gulo*
Wood thrush	*Hylocichla mustelina*
Wood turtle	*Glyptemys insculpta*

Y

Yellow birch	*Betula alleghaniensis*
Yellow-rumped warbler	*Setophaga coronata*

Index

G

H

J

O

P

R